AquaRating.

An International Standard for Assessing Water and Wastewater Services

Matthias Krause, Enrique Cabrera Rochera, Francisco Cubillo, Carlos Díaz, Jorge Ducci.

Published by   **IWA Publishing**
                      Alliance House
                      12 Caxton Street
                      London SW1H 0QS, UK
                      Telephone: +44 (0)20 7654 5500
                      Fax: +44 (0)20 7654 5555
                      Email: publications@iwap.co.uk
                      Web: www.iwapublishing.com

First published 2015

The publisher makes no representation, express or implied, with regard to the accuracy of the information contained in this book and cannot accept any legal responsibility or liability for errors or omissions that may be made.

Disclaimer
The information provided and the opinions given in this publication are not necessarily those of IWA and should not be acted upon without independent consideration and professional advice. IWA and the Editors and Authors will not accept responsibility for any loss or damage suffered by any person acting or refraining from acting upon any material contained in this publication.

Published by: IWA Publishing

ISBN: 9781780407395 (Paperback)
ISBN: 9781780407401 (eBook)

Contents

Chapter 2
PE Investment Planning and Implementation Efficiency....69

Chapter 3

Acknowledgements

AquaRating is the result of several years of work and has benefited from the contributions of many people. Among all those people to whom we, the authors and members of the AquaRating Technical Committee, wish to thank explicitly, we want to highlight the members of the broader AquaRating team who have worked on a daily basis on the project and whose inputs have been critical to render AquaRating what it is today: María del Rosario Navia, Daniel Fernández, Iván Montalvo and Raimon Puigjaner.

We are grateful to the participants of the several focus groups and seminars held on AquaRating, who inverted their valuable time to share their opinions and give recommendations, this way contributing to the development of the present rating standards.

The thirteen utilities who participated in the testing of AquaRating's pilot version in nine countries of Europe and Latin America deserve a very special thank you, for the effort they dedicated to the testing process and for their recommendations that have resulted in enhancements of the AquaRating standards. These utitlies are, in alphabetical order: Agua y Saneamientos Argentinos (AySA), Argentina; Aguas Andinas, Santiago, Chile; Aguas de Alicante, Spain; Aguas de Cartagena, Colombia; Companhia de Saneamento Básico do Estado de São Paulo (SABESP), Brazil; Corporación del Acueducto y Alcantarillado de Santiago, Dominican Republic; Empresa de Acueducto y Alcantarillado de Pereira, Colombia; Empresa Municipal de Aguas de Córdoba, Spain; Empresa Pública Metropolitana de Agua Potable y Saneamiento de Quito, Ecuador; Empresas Públicas de Medellín (EPM), Colombia; FCC Aqualia Almería, Spain; Obras Sanitarias del Estado, Uruguay; Servicios de Agua y Drenaje de Monterrey, Mexico.

Last but no least we want to thank for the indispensable institutional support of the Inter-American Development Bank (IDB) and the International Water Association (IWA) and their management staff, in particular Federico Basañes, Sergio I. Campos G. and Paul Constance for IDB and Ger Bergkamp and Tom Williams for IWA. IDB from the very beginning has promoted and financed AquaRating's development and has dedicated a technical and support team to it. IWA, as the knowledge network of the water sector, has significantly contributed to AquaRating's development and dissemination and is leading its implementation into practice.

March 2015

AquaRating Technical Committee
Matthias Krause (coordinator)
Enrique Cabrera Rochera
Francisco Cubillo
Carlos Díaz
Jorge Ducci

About the authors

Dr. Matthias Krause
Coordinator of the AquaRating Technical Committee, at present is a Senior Specialist at the Water and Sanitation Division of the Inter-American Development Bank (IDB) in Washington, D.C. Before joining the IDB in 2010, he worked for eight years as a senior researcher of the German Development Institute (DIE) in Bonn, where he advised the German government and international organizations in questions related to infrastructure for development. His areas of specialization are water governance, private sector participation, corporate governance and performance assessment of water and sanitation utilities. He is author of the book "The political economy of water and sanitation". Matthias Krause holds a PhD in Economics by the University Justus-Liebig of Gießen, Germany.

Dr. Enrique Cabrera Rochera
Member of the AquaRating Technical Committee, has more than 15 years of experience in the field of management of urban water services. His PhD thesis analyzed the topics of performance assessment and benchmarking. He is co-author of the International Water Association's (IWA) Manuals on Performance Indicators for Water Supply Services and Benchmarking. He currently is the President of the Board of Directors of IWA Publishing and a Member of IWA's Board of Directors, as well as the Chair of the IWA Benchmarking and Performance Assessment Specialist Group. He has additionally played a key role in the drafting of the ISO 24500 standards on water services, where he acted as President of Working Group 2 (which drafted the ISO 24510 standard). Enrique Cabrera Rochera is Associate Professor at the Polytechnic University of Valencia where he

lectures on fluid mechanics, and he currently serves as the Deputy Director of Innovation and Communication of the Industrial Engineering Faculty.

Francisco Cubillo

Member of the AquaRating Technical Committee, is the Deputy Director of Investigation, Development and Innovation at Canal Isabel II Gestión S.A., the company responsible for water supply, distribution and wastewater treatment of more than 6 million inhabitants in 170 municipalities in the Madrid area (Spain). He is co-author of the International Water Association's (IWA) Manual on Performance Indicators for Water Supply Services. Between 2001 and 2011 he was the chair of IWA's Urban Water Efficient Management Specialist Group, and currently chairs IWA's Cluster of Alternative Resources as well as the Investigation, Development and Innovation Committee of the Spanish Association of Water and Sanitation Providers (AEAS). Francisco Cubillo has more than 30 years of experience in managing urban water services and water resource planning, both in public entities and private engineering consulting firms. He has published 22 books and more than 100 technical papers, as well as been a lecturer of a wide array of courses on supply management systems, hydrology, technology development and environment.

Carlos Díaz

Member of the AquaRating Technical Committee, has more than 30 years of professional experience both in consulting and corporate management. He holds degrees in commercial engineering, with a specialization in business administration, and in accounting, and currently works as a consultant and as the Managing Director of Soluciones Integrales S.A., a consulting company specialized in economic, financial and technical analyses of infrastructure, in particular water and sanitation. He has worked as a consultant for the Inter-American Development Bank (IDB) in the field of water and sanitation, as a board member of two Chilean water and sanitation companies, and, during his early professional career, as a financial specialist at Empresa Metropolitana de Obras Sanitarias (EMOS; nowadays "Aguas Andinas") in Santiago, Chile. Between April 1990 and July 1998 Carlos Díaz was the National Director for Accounting and Finance of the Chilean Ministry of Public Works, and from October 1979 to March 1990 he worked as consultant, lead consultant, and director of consulting at Price Waterhouse Chile's Consulting Department.

Jorge Ducci

Member of the AquaRating Technical Committee and economist by the Catholic University of Chile and Cornell University, currently is a Lead Specialist at the Water and Sanitation Division of the Inter-American Development Bank (IDB) in Washington, D.C. Between 1983 and 1990 he worked as an economist in project appraisal for IDB's Infrastructure Division, and from 1990 to 1993 he was the National Director for Planning of the Chilean Ministry of Public Works. Since then until 2008 he was partner and Managing Director of Soluciones Integrales S.A., a consulting company specialized in economic, financial and technical analyses related to infrastructure, in particular water and sanitation, urban areas, environment, public service concessions, and socioeconomic project evaluation. Jorge Ducci is the author of several studies and papers on water issues, among others of an analysis of the reasons why foreign operators left the water supply business in Latin America ("Salida de operadores privados internacionales de agua en América Latina").

Introduction

WHAT IS AQUARATING

AquaRating is a new rating system for the water sector. AquaRating's objective is to facilitate continual improvement of drinking water and wastewater services by providing rigorous, systematic and universal assessment.

AquaRating provides a new universal standard by which to assess drinking water and wastewater utilities. The AquaRating standard comprehensively assesses drinking water and wastewater services based on 112 assessment elements organized into 8 areas, each of which is assigned a rating. These ratings are then aggregated into a single rating (from 0 to 100) for the utility. The assessment elements consist of good practices, performance indicators and information quality. Total compliance with practices and attainment of the most demanding indicator scores equates as delivery of excellent service and, therefore, earns a maximum rating of 100 points.

The rating system is managed by an independent organization - which operates as the AquaRating agency - reporting to the International Water Association (IWA). The system has been developed as an initiative of the Inter-American Development Bank (IDB) in close collaboration with the IWA.

For further information about AquaRating, see www.aquarating.com

THE BENEFITS OF AQUARATING

The AquaRating standard has been designed to assess drinking water and wastewater utilities operating mainly in urban areas and located anywhere in the world. The main users of AquaRating are envisaged to be drinking water and wastewater utilities. Nevertheless, the system is also expected to be useful for other water sector stakeholders, such as financial institutions, development co-operation agencies, regulators and public authorities as, based on a transparent and public rating structure, it objectively assesses the main management elements of the services delivered by a utility within its allotted area.

AquaRating is expected to produce the following main benefits:

- Creation of an objective, independent and universal system by which to assess drinking water and wastewater utilities.
- Provision of complete and detailed analysis of utility performance.
- Identification of opportunities to improve utility performance, elaboration and monitoring of improvement plans.
- Establishment of utility reputation for performance and improvement.
- Improvement of access to financial and capacity-building resources offered by public authorities and financial institutions.
- Improvement of access to new markets for utilities rated highly for their performance and commitment to improving.

Utilities that are fully aware of their current standard of performance and that draw up and implement sound improvement plans will be likely to deliver services which are superior in terms of quality, efficiency, sustainability and transparency.

THE AQUARATING PROCESS

AquaRating rates the utility and produces a comprehensive report that identifies opportunities for improvement by providing disaggregated information on the factors used to determine the rating and the scores achieved for each of them. This allows utilities to establish action plans for the various areas for improvement identified in the rating process.

Full application of the AquaRating process comprises the following stages:

Stage 1: Performance assessment
In this stage, the utility provides the information needed to perform self-assessment against the AquaRating standards.

Stage 2: Performance certification
In this stage, an auditor verifies the information provided by the utility and, based on that audit, AquaRating certifies the performance rating obtained.

Stage 3: Performance improvement
In this stage, the utility draws up an action plan based on the performance assessment carried out by AquaRating.

In the first stage, the utility provides all the information needed to perform self-assessment of its performance against the rating standards. This process is managed via a secure IT platform to which only the utility and AquaRating have access. Once it has submitted its data, the utility can generate a variety of internal reports on either overall performance or particular areas, allowing it both to gain an overview and to analyze individual practices.

Certification of the rating is a critical component of the AquaRating process, as it gives the rating objectivity and universality. To guarantee these two aspects, and to establish the reliability of the information supplied, all information leading to calculation of the final rating must be supported by documentary evidence and validated by an independent audit. This audit is carried out by auditors accredited by AquaRating. In this stage, the accredited auditor selected by the utility is given access to the data on the IT platform and audits that information. The outcome of the certification stage is issue of a certified report and the AquaRating quality seal.

It is important to emphasize that the rating and the information compiled in the assessment process is kept confidential. The decision to publish all or part of that information is entirely at the utility's discretion.

THE AQUARATING STANDARD

AquaRating's characteristics make it a universal standard for assessing drinking water and wastewater utilities. These characteristics are as follows:

- Universality (validity in any context)
- Comprehensive assessment (covering all areas relevant to performance)
- Guarantee of a complete utility rating based on indicators and practices
- Ability to assess current performance and improvement potential
- Ability to provide relevant information for the improvement of services
- Auditability

AquaRating consists of two basic assessment elements:

1) Indicators: Based on IWA-established guidelines and ISO 24500 standards, AquaRating indicators are accompanied by a normalization function that establishes the rating for each particular element.

2) Good practices: For each item to be assessed, AquaRating presents a collection of good practices that characterize excellent service performance. The utility's rating depends on the number of practices implemented in the assessed service.

By assessing the way in which utilities manage services as well as the numerical outcomes of this management, geographical, economic and social differences do not have as great an influence on the assessment as they would under a purely indicator-based approach. Moreover, by using practices as assessment elements, it is possible to establish projections for the utility: a utility currently performing well but lacking good practices may face sustainability problems, while a utility performing poorly but with good management practices in place will certainly improve in the future. Finally, one of the virtues of the good practices included in AquaRating is that they in themselves provide guidelines for improvement, since practices not implemented in the service may be established as objectives for the utility.

Another critical part of AquaRating is its audit system and guaranteed information quality. Data quality has been recognized in many indicator systems as a cornerstone for validating assessment in a sector in which, because of its nature, information reliability may be highly variable. Good quality information in a drinking water and wastewater utility is not fortuitous, but the result of planning and of adequate data management. Hence, AquaRating assesses the reliability of supporting information through a system of reliability tables and modifies the rating as a function of that reliability.

Finally, the IT application used by the rating system adapts easily to scope of services being offered by the utility to be rated. AquaRating defines the following service stages:

- Drinking water production
- Drinking water distribution
- Wastewater collection
- Final wastewater disposal and treatment

It also defines the following service functions:

- Customer management
- Operation and maintenance of infrastructure linked to the service stages mentioned above
- Financing of replacement of existing physical assets
- Financing of augmentation or expansion of existing or new physical assets

Should the utility not have the mandate to perform one or more of the service stages, the system applies a series of filters that ignore those assessment elements not applicable to the utility and recalculates the assessment element weightings accordingly. The minimum requirement to apply AquaRating consistently is that the utility's mandate includes the functions of customer management and operation and maintenance of - at the very least - drinking water distribution or wastewater collection. The standards presented here describe the full system configuration as applied to a utility with a mandate to perform all the service stages mentioned above.

AQUARATING STANDARD STRUCTURE

AquaRating is divided into 8 rating areas: Service Quality, Investment Planning and Implementation Efficiency, Operating Efficiency, Business Management Efficiency, Financial Sustainability, Access to Service, Corporate Governance, and Environmental Sustainability.

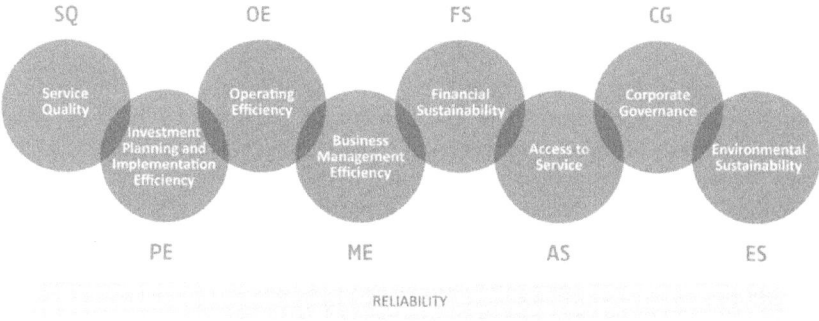

Fig. 1. AquaRating rating areas

Each rating area is divided into sub-areas. Thus, for example, the 'Service Quality' area is divided in 4 sub-areas, as shown in the figure below.

Fig. 2. Service quality hierarchy

The assessment elements necessary for rating can be found in each of the aforementioned sub-areas. Each sub-area may contain elements comprising good practices, indicators or both. The system assesses the reliability of the supporting data for each of the assessment elements and using reliability tables modifies the element's rating according to that reliability by means of a correction factor.

The ratings for each of the areas are obtained by weighted aggregation of all the elements hierarchically below that area. Thus, starting from the assessment elements, each level of the AquaRating system is assigned a rating from 0 to 100. These ratings are weighted within the same level using previously determined weights. Each area and sub-area gets a rating, from 0 to 100, resulting from the combination of those elements.

To get a rating for each element from 0 to 100, it is necessary to normalize the results of each individual assessment. In the case of lists of good practices, each of those practices has a relative importance in relation to the rest (weight). If all practices are complied with, that supposes a rating of 100 for that assessment element, while any partial compliance will result in a proportional reduction of the rating as a function of the weight of the practice not complied with.

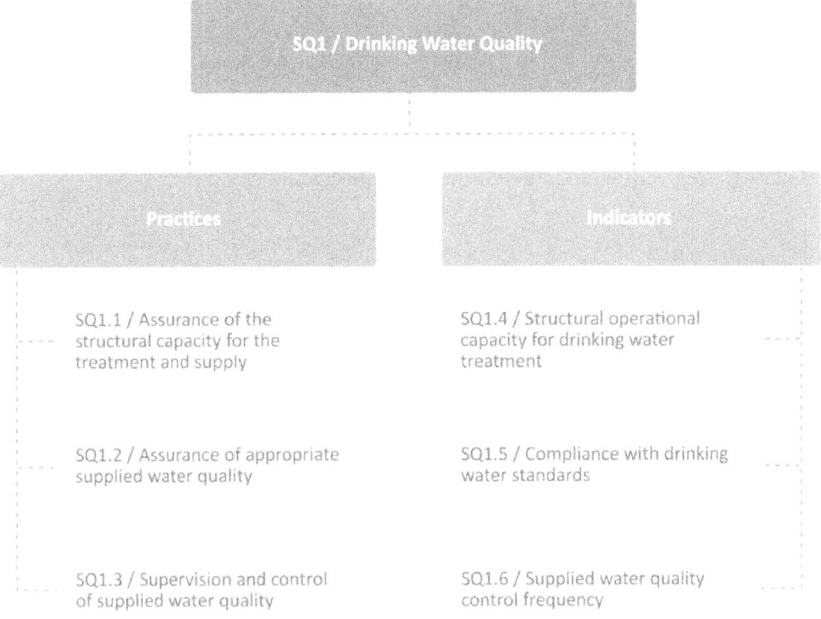

Fig. 3. Assessment elements for the SQ1 Drinking Water Quality sub-area

Practices	Reliability	Weight
1	Disinfection equipment exists and covers the entire population served. In cases in which water sources are used that, in accordance with "applicable standards", do not require treatment, the population supplied from these sources will be considered as meeting this criterion.	T. 1	3
2	Drinking "water treatment" facilities exist (with a nominal capacity greater than or equal to maximum daily consumption), and cover the entire population served, to treat water requiring more than disinfection to comply with regulations.	T. 8	2
3	Alternative "treatment" facilities exist to provide drinking water in hydraulically independent zones with populations of over 100,000 inhabitants. At least 50% of the population of such zones will have more than one source of drinking water. In "systems" in which these circumstances do not apply, this practice is considered complied with at maximum reliability level.	T. 8	1
4	Analyses are carried out to identify "zones at risk of not complying" with drinking water quality standards, and, if this is the case, appropriate actions are identified.	T. 4	1
5	Distribution network design criteria exist that consider water quality issues (such as time that water remains in the network, removal of blanked pipe sections, etc.).	T. 2	1
6	Studies are carried out to identify zones in which disinfectant concentrations are at risk of falling below levels established in applicable regulations and to define measures to ensure minimum and more homogeneous disinfectant concentration in distribution networks.	T. 4	1

Fig. 4. List of good practices for SQ1.1 Assurance of structural capacity for treatment and supply

In the case of indicators, normalization is accomplished by means of a normalization function which defines the desired values for each indicator.

Introduction

Said functions present the indicator value in the abscissa axis and the normalized AquaRating value in the ordinate axis. Figure 5 shows how the percentage of compliance with drinking water standards (% of compliant samples) yields a normalized value of 0 for 80% sample compliance. 90% compliance yields a normalized value under 30, whereas 100% compliance yields the maximum score of 100.

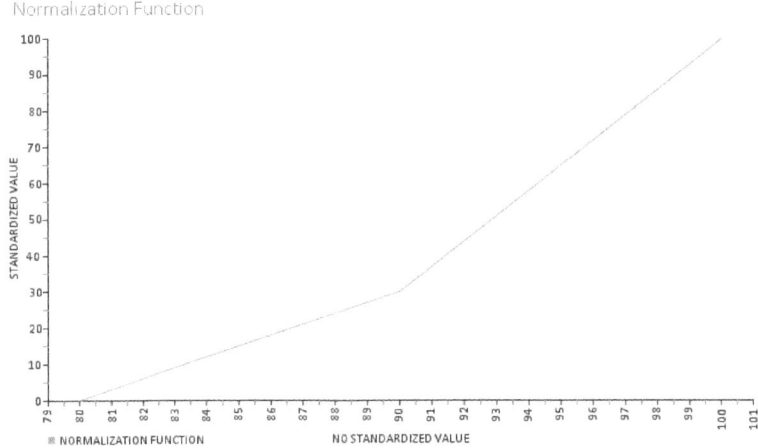

Fig. 5. Normalization function for SQ1.5 Compliance with drinking water standards

As commented, aggregation of ratings at each level is performed by adding together the weighted ratings of all elements in that level. Because of this, each rated element has a unique pre-assigned weight in the system. Figure 6 shows how the various elements in a level determine the successive rating of sub-areas, areas and overall service.

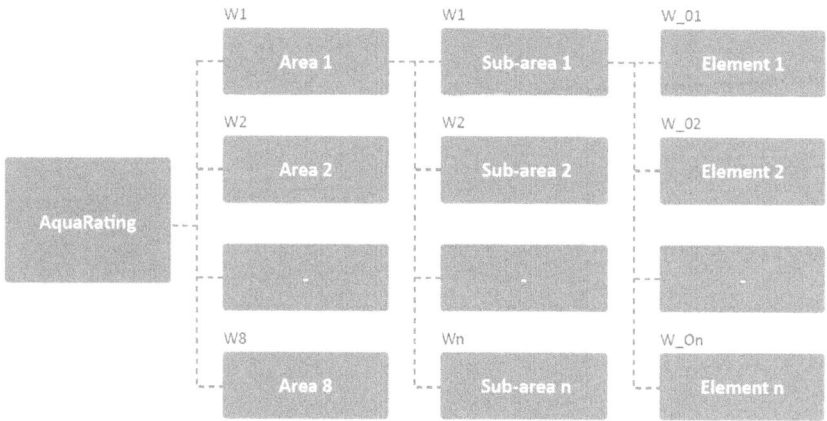

Fig. 6. Weighted aggregation of the various ratings by level to produce the overall rating

PRACTICAL ASPECTS

This document contains a detailed definition of the AquaRating standard in its full configuration and all the information needed to assess and understand the rating system. It includes the complete hierarchy of areas and sub-areas, as well as a detailed definition of each assessment element and the weights assigned at the assessment element, sub-area and area levels. It also contains a glossary which includes key terms for interpreting the assessment elements. These terms are shown in the document in quotation marks.

The rest of the document is structured as follows: the main body of the text defines in detail each of the eight assessment areas while the annexes include a complete list of reliability tables, a glossary, and a list describing the system's weighting structure.

SQ Service Quality

The quality with which drinking water and wastewater services are delivered is undoubtedly the most conclusive element among those allowing assessment of utility management. This quality has a direct impact on users, their health, their comfort and their perception of the utility. It also reflects the consequences of current operating and management practices and, especially, of those carried out previously and which have made it possible to achieve current quality levels.

Assessment has been divided into four rating sub-areas, which consider aspects concerning quality of water delivered for consumption, delivery quantity and continuity parameters, conditions in which wastewater is collected and transported from properties and businesses, and forms of user service, including user perceptions of the set of services received.

The wastewater treatment service has been included as one of the rating criteria for the Environmental Sustainability area, as it is in relation to the environment that it is best integrated into the AquaRating system. Grouping it there does not detract from its nature as a service delivered and often billed separately and which, consequently, may be carried out according to certain standards and with differing degrees of efficiency.

Assessment of the Service Quality area is limited to conditions and results in terms directly linked to the service delivered, i.e. the efficacy with which the services are delivered. Hence, unlike other assessment areas, it does not incorporate considerations relative to the degree of efficiency in delivery of those services. However, it does include certain aspects indirectly related to planning or operating practices, such as structural capacity or of operation and control of equipment, facilities or processes.

The sub-areas of evaluation are:

SQ1 Drinking water quality
SQ2 Distribution of drinking water for use and consumption
SQ3 Wastewater collection
SQ4 User service

SQ1 Drinking water quality

The quality of the service delivered in terms of the potability of the water delivered for use and consumption is assessed according to 3 indicators and 3 groups of applied practices and processes. The indicators assess the degree of coverage of the existing operational infrastructure used to treat raw water and transform it into water fit for consumption, the degree of compliance with standards relative to water for consumption applicable in each case, and the intensity with which quality control is performed on the water delivered. As far as the assessment elements related to the practices are concerned, these consider procedures intended to ensure that sufficient structural capacity exists to perform appropriate treatment, operation and control.

Practices

SQ1.1   Assurance of structural capacity for treatment and supply
SQ1.2   Assurance of appropriate supplied water quality
SQ1.3   Supervision and control of supplied water quality

Indicators

SQ1.4   Structural operational capacity for drinking "water treatment"
SQ1.5   Compliance with drinking water standards
SQ1.6   Supplied water quality control frequency

SQ1.1 Assurance of structural capacity for treatment and supply

Type: Best Practices
Service: Drinking Water
Normalization: Weighted by practices
Glossary: Applicable regulations, System, Conventional water treatment, Zones at risk of non-compliance with drinking water quality standards
Definition: Includes:

	Practices	Reliability	Weight
1	Disinfection equipment exists and covers the entire population served. In cases in which water sources are used that, in accordance with "applicable standards", do not require treatment, the population supplied from these sources will be considered as meeting this criterion.	T. 1	3
2	Drinking "water treatment" facilities exist (with a nominal capacity greater than or equal to maximum daily consumption), and cover the entire population served, to treat water requiring more than disinfection to comply with regulations.	T. 8	2
3	Alternative "treatment" facilities exist to provide drinking water in hydraulically independent zones with populations of over 100,000 inhabitants. At least 50% of the population of such zones will have more than one source of drinking water. In "systems" in which these circumstances do not apply, this practice is considered complied with at maximum reliability level.	T. 8	1
4	Analyses are carried out to identify "zones at risk of not complying" with drinking water quality standards, and, if this is the case, appropriate actions are identified.	T. 4	1
5	Distribution network design criteria exist that consider water quality issues (such as time that water remains in the network, removal of blanked pipe sections, etc.).	T. 2	1

Studies are carried out to identify zones in which disinfectant concentrations are at risk of falling below levels established in applicable regulations and to define measures to ensure minimum and more homogeneous disinfectant concentration

6 in distribution networks.   T. 4   1

SQ1.2 Assurance of appropriate supplied water quality

Type: Best Practices
Service: Drinking Water
Normalization: Weighted by practices
Glossary: Applicable regulations, System, Contingency, Preventive maintenance, Corrective maintenance, Preventive maintenance protocol, Corrective maintenance protocol
Definition: Includes:

	Practices	Reliability	Weight
1	Protection measures exist at all raw water intakes (signage, perimeter protection, fencing, etc.) for sources of raw water incorporated into the "system" rated.	T. 4	1
2	"Protocols" exist to perform "preventive maintenance" on treatment plants, as do corresponding maintenance records.	T. 2	1
3	"Protocols" exist to perform "corrective maintenance" on treatment plants, as do corresponding maintenance records.	T. 2	1
4	Automated processes exist in water treatment plants serving more than 5,000 inhabitants to ensure operation in the absence of personnel, or personnel are on duty 24/7 if automated processes do not exist.	T. 5	3
5	Protocols exist to analyze and resolve non-compliance with "applicable regulations" regarding water quality and notify the competent authority.	T. 2	2
6	Safety plans exist for "contingencies" regarding water quality.	T. 2	1
7	In "systems" with various alternative supply sources, protocols exist to assure water quality on initial use of new supply sources (wells or surfacewater); otherwise, this practice is considered complied with at maximum reliability level.	T. 120	1
8	Protocols exist to assure water quality when integrating new infrastructure.	T. 120	1

SQ1.3 Supervision and control of supplied water quality

Type: Best Practices
Service: Drinking Water
Normalization: Weighted by practices
Glossary: Applicable regulations, Corrective maintenance
Definition: Includes:

	Practices	Reliability	Weight
1	Self-applied supplied water quality control protocols exist, as do records of the findings, applying criteria at least as stringent as those set by "applicable regulations".	T. 6	1
2	Laboratories that carry out the analyses (whether own or external) hold ISO 17025 certification.	T. 2	2
3	Operational equipment measuring physical and chemical parameters is available in all water treatment plants (either as permanently installed equipment or permitting sample-taking at inlet and outlet and in intermediate processes).	T. 1	2
4	Records are kept of operation parameters measured in all treatment plants.	T. 6	2
5	Alarm thresholds exist for "corrective maintenance" and operation adjustment.	T. 2	3
6	Remote control systems are available to manage processes and internal parameters in treatment plants.	T. 3	1
7	Automatic water quality monitoring stations are available (in at least 50% of the zone supplied) at the outlets of the treatment plants or tanks.	T. 1	1
8	A network of stationary facilities exists to facilitate collection of water quality samples with a representativeness of at least 1 per 20,000 inhabitants as per census population.	T. 8	2

SQ1.4 Structural operational capacity for drinking "water treatment"

Reflects the extent of coverage of facilities used to treat water for use and consumption. It is assessed by the percentage of population capable of receiving treated water via the treatment and distribution infrastructure in operation, regardless of treatment type and operational efficacy. Even though there will be cases in which the availability of this infrastructure will not depend directly on the entity which operates the "system", it is considered a relevant indicator for estimating the potential quality of the service (linked to water quality) that "system" users receive.

Definition: Percentage of population in the "geographical area to be rated" for drinking water supply served by supply eligible for "treatment" in a facility with sufficient nominal capacity and the corresponding distribution network.
Type: Indicator
Service: Drinking Water
Glossary: System, Conventional water treatment, Geographical area to be rated
Formula: ([CS1-V1]/[CS1-V2])*100   Unit: %
Normalization Function:

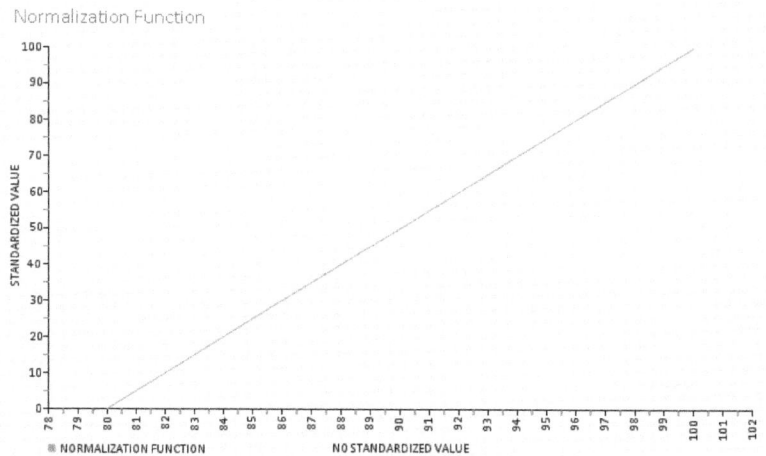

Variables

[CS1-V1] Population served by drinking water supply eligible for "treatment" of some kind.
Definition: Population provided with a household connection in the "geographical area to be rated" for drinking water supply and served by supply susceptible to "treatment" of some kind in a facility with sufficient nominal capacity (at the end of the calendar year preceding the rating date).
Units: inhabitants
Reliability: Table 7

[CS1-V2] Population with a household connection in the "geographical area to be rated" for drinking water supply.
Definition: Population with a household connection in the "geographical area to be rated" for drinking water supply (at the end of the calendar year preceding the rating date).
Units: inhabitants
Reliability: Table 100

SQ1.5 Compliance with drinking water standards

Reflects the standard of water quality supplied in the calendar year preceding the rating date. It is assessed by comparing water sample quality analyses against applicable standards and is quantified by the percentage of inhabitants served who receive water that meets the conditions established in the "applicable regulations".

Definition: Percentage of inhabitants whose water supply complies with "applicable standards" as a proportion of total inhabitants in the "geographical area to be rated" for drinking water supply in the full calendar year preceding the rating date.
Type: Indicator
Service: Drinking Water
Glossary: Applicable regulations, Geographical area to be rated
Formula: ([CS1-V3]/[CS1-V2])*100   Unit: %
Normalization Function:

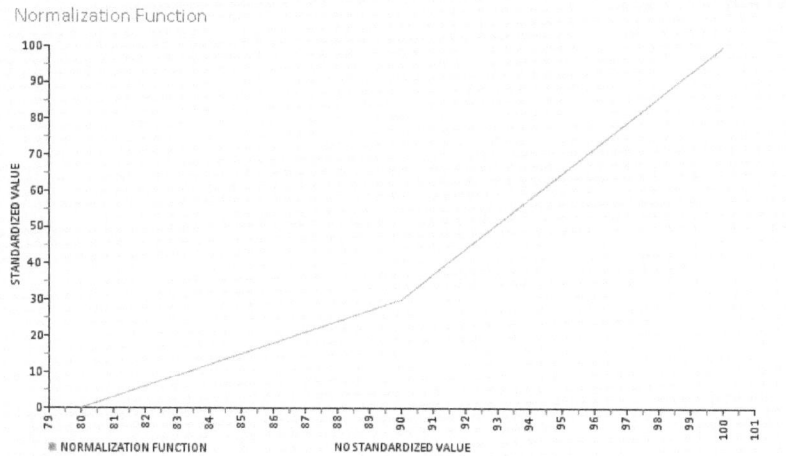

Variables

[CS1-V2] Population with a household connection in the "geographical area to be rated" for drinking water supply.

Definition: Population with a household connection in the "geographical area to be rated" for drinking water supply (at the end of the calendar year preceding the rating date).

Units: inhabitants

Reliability: Table 100

[CS1-V3] Number of inhabitants for whom "applicable water sample quality standards" have been complied with.

Definition: Number of inhabitants for whom "applicable water sample quality standards" have been complied with throughout the full calendar year preceding the rating date. It is assumed that standards have been complied with in a zone when the controls established in the applicable standards have been carried out and the criteria established have been met in full.

Units: inhabitants

Reliability: Table 9

SQ1.6 Supplied water quality control frequency

Definition: Percentage of days per year on which "samples representative of supplied quality" within the entire "system" are taken and analyzed.
Type: Indicator
Service: Drinking Water
Glossary: System, Representative sample of supplied quality
Formula: ([CS1-V4]/365)*100   Unit: %
Normalization Function:

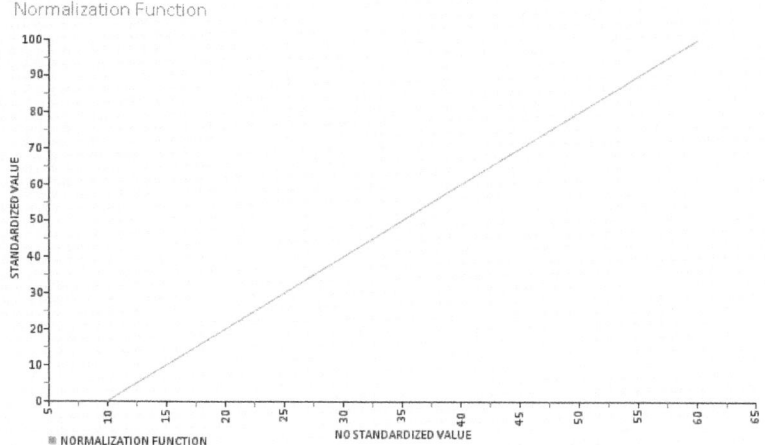

Variables

[CS1-V4] Number of days per year on which "representative samples of supplied quality" are taken and analyzed.

Definition: Number of days per year on which "representative samples of supplied quality" are taken and analyzed throughout the full calendar year preceding the rating date.

Units: days

Reliability: Table 10

SQ2 Distribution of drinking water for use and consumption

The service quality rating linked to drinking water distribution considered in this section is limited to those aspects of the service that depend on proper operation of the intake, transportation, tank storage and distribution "systems". However, it does include issues related to water quality (SQ1), user service (SQ4) and access to service (AS). Therefore, the area of consideration is only the area covered by supply infrastructure to properties or points of use and consumption managed by the utility that is rated within the system.

The quality of the service delivered will depend, among other factors, on the capacity of existing infrastructure to exploit the water resources available and to transport and distribute them after treatment. This capacity is in turn a consequence of appropriate planning, design, implementation and investment policy by authorities with the power and obligation to extend, enlarge and renew supply and distribution infrastructure.

This group of assessment elements considers everything related to continuity of water supply for consumption as regards compliance with required basic water parameters. The time it takes to connect new users to the supply "system" is also considered, understanding the delay between requesting a connection and the connection becoming operational to be a specific form of discontinuity.

Some parameters usually used to estimate service continuity, such as the number of ruptures in pipelines, have been included in the Operating Efficiency area (OE), as it is considered more appropriate to link them to infrastructure management practices even though they could be used to determine discontinuities in supply and service conditions.

Practices

SQ2.1   Assurance of structural capacity for supply and distribution
SQ2.2   Assurance of supply continuity during operation
SQ2.3   Supervision and control of supply continuity

Indicators

SQ2.4   Supply continuity

SQ2.5   Time taken to connect new users to the drinking water service

SQ2.1 Assurance of structural capacity for supply and distribution

Type: Best Practices
Service: Drinking Water
Normalization: Weighted by practices
Glossary: System, Contingency
Definition: Includes:

Practices		Reliability	Weight
1	Standard service pressure and continuity values for water supply and distribution are accepted and applied.	T. 2	3
2	Water supply and distribution infrastructure design is conceived to minimize impact due to "contingencies" and to comply with service standards.	T. 2	2
3	Plans to renew supply and distribution "system" elements adopt criteria that take into account the risk of impact on service continuity.	T. 2	2
4	Planning and adjustment of water supply and distribution infrastructure adopt criteria to prevent risk of service interruption and unintended variations in pressure.	T. 2	1

SQ2.2 Assurance of supply continuity during operation

Type: Best Practices
Service: Drinking Water
Normalization: Weighted by practices
Glossary: System
Definition: Includes:

Practices	Reliability	Weight
1	Visual inspection of supply and distribution system elements and accessibility to them is carried out at least once every 3 years.	T. 11	1
2	Verification of operational status of supply and distribution system elements for which visual inspection or operational test is possible is carried out at least once every 3 years.	T. 11	1
3	Supply and distribution "system" elements are differentiated and those of fundamental or strategic importance are highlighted. Such differentiation must be stated in all infrastructure databases used for planning and operation and must be indicated on-site on all visible or maneuverable system elements.	T. 4	2
4	Supply and distribution "system" elements of fundamental or strategic importance are checked at least every 6 months.	T. 6	2
5	Problems detected in supply and distribution "system" facility inspections, including inconsistencies with the information available, are resolved within 6 months.	T. 6	2
6	Systematic campaigns to detect leaks and hidden ruptures are carried out on at least 5% of network length every year.	T. 6	1

SQ2.3 Supervision and control of supply continuity

Type: Best Practices
Service: Drinking Water
Normalization: Weighted by practices
Glossary: System, Contingency
Definition: Includes:

Practices	Reliability	Weight
1 Specific human and material resources are available 24/7 to manage "contingencies" in water supply and distribution "systems".	T. 5	3
2 GIS tools are available to support isolation, repair and resolution of "contingencies" in supply and distribution "systems".	T. 3	3
3 Early warning mechanisms are available 24/7 (remote control and receipt of warnings referring to supply and distribution "systems").	T. 3	3

SQ2.4 Supply continuity

This assessment element evaluates service continuity based on the number of hours during which the "hydraulic conditions" at the point of connection to the distribution "system" at each "property" supplied are sufficient for use and consumption.

Definition: Number of hours during which "hydraulic conditions for use and consumption" have not been met for each "property" in the full calendar year preceding the rating date.
Type: Indicator
Service: Drinking Water
Glossary: System, Sufficient hydraulic conditions for use and consumption, Property
Formula: [CS2-V1]/[CS2-V2]   Unit: hours
Normalization Function:

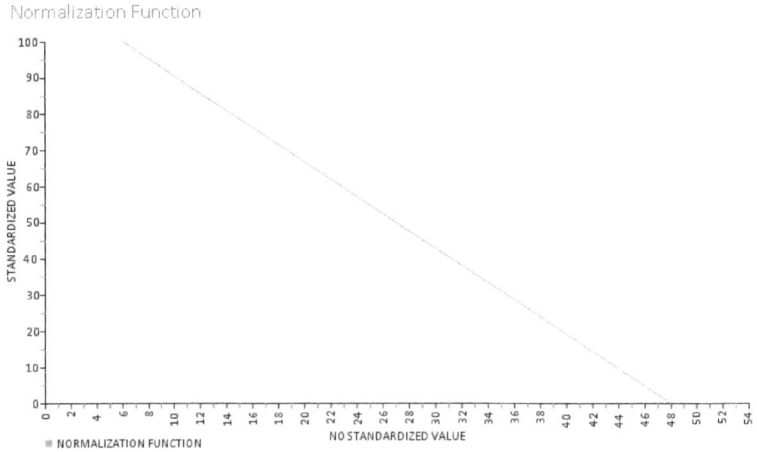

Normalization Function

Variables

[CS2-V1] Total number of hours of interrupted supply
Definition: Total number of hours of interrupted supply or without the necessary "hydraulic conditions for supply and consumption" in each property throughout the last full calendar year. In "systems" in which continuous supply is generally not available, interruption hours will be applied to all properties located in the area that lacks such service.
Units: hours
Reliability: Table 12

[CS2-V2] Number of "properties" supplied
Definition: Number of "properties" supplied at the end of the last full calendar year.
Units: properties
Reliability: Table 13

SQ2.5 Time taken to connect new users to the drinking water service

Assesses the average time taken to connect a new user to the supply "system". Application times are not considered, as it is assumed that in principle these will be short and will be assessed in the user service section. Likewise, delays attributable to issue of permits required to implement the works needed, or attributable to other processes for which the utility is not exclusively responsible, are not considered.

Definition: Average time elapsed between completing application for the service and "completion" of connection of the new user's property to the network.

Type: Indicator

Service: Drinking Water

Glossary: System, Finished works

Formula: [CS2-V3]   Unit: days/connections

Normalization Function:

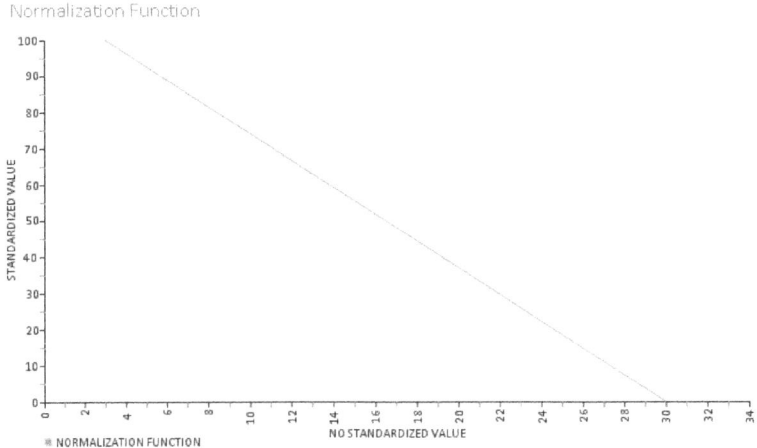

Variables

[CS2-V3] Average time elapsed between completing application for the service and "completion" and notification of connection to the new user's property.
Definition: Average time elapsed between completing application for the service and "completion" and notification of connection to the new user's property (for all connections made in the last full calendar year).
Units: days
Reliability: Table 14

SQ3 Wastewater collection

This sub-area considers the general wastewater collection "system" from the point of connection to private "systems" to delivery to wastewater treatment plants or return to the environment. The service considered is collection of wastewater from all activities that discharge wastewater into the "system".

Although AquaRating is not applicable to services that exclusively perform rainwater drainage, it inevitably considers the rainwater drainage function in conjunction with wastewater collection in cases in which providers perform this function as an integral or combined, direct or indirect, or joint or separate part of the service.

Also included as an assessment element is the time taken to connect physically to the sewerage system, which is equivalent to the time taken to connect materially to the supply network.

From the point of view of service quality, it is difficult to determine which incidents have an impact on sewerage "system" users. In fact, except for incidents that block individual wastewater discharge connections, the most common occurrence is blockage of elements and zones of the sewerage network that become evident on public thoroughfares. These have an impact but do not result in clear and immediate interruption of service to properties.

Assessment is limited to incidents attributable to the "system's" state or operation, excluding those caused by third parties such as theft of manhole covers or similar. Although in many cases these incidents represent a serious maintenance problem, they do not create operational problems affecting collection and transportation of wastewater and rainwater.

Practices

SQ3.1   Assurance of structural capacity for wastewater collection
SQ3.2   Assurance of wastewater collection from operation
SQ3.3   Supervision and control of the wastewater collection service

Indicators

SQ3.4   Time taken to resolve "incidents" in the wastewater collection network

SQ3.5   Time taken to connect to the wastewater service
SQ3.6   Stormweather "incidents"

SQ3.1 Assurance of structural capacity for wastewater collection

Type: Best Practices
Service: Sanitation
Normalization: Weighted by practices
Glossary:
Definition: Includes:

	Practices	Reliability	Weight
1	An up-to-date wastewater plan is in force that includes the wastewater collection network, or a specific wastewater collection network plan exists.	T. 6	1
2	The objectives and timetables of the wastewater plan or specific wastewater collection network plan are fulfilled.	T. 6	1
3	Standards for construction elements employed in the wastewater collection network and for implementation of connections to this network by entities other than the utility exist and are applied.	T. 6	1
4	Standards for use of the wastewater collection network by other services, such as fiber optics, exist and are applied.	T. 6	1

SQ3.2 Assurance of wastewater collection from operation

Type: Best Practices
Service: Sanitation
Normalization: Weighted by practices
Glossary: System, Preventive maintenance, Preventive maintenance protocol
Definition: Includes:

Practices	Reliability	Weight
1	"Protocols" and records of "preventive maintenance" performed on the wastewater collection "system" exist (inspection frequency as a function of parameters such as pipe age, etc.).	T. 6	1
2	A computerized system to manage "preventive maintenance" of the wastewater collection "system" exists and is used.	T. 3	1
3	Specific equipment is available to permit inspection of difficult-to-access zones of the wastewater collection network.	T. 1	1
4	Collector cleaning protocols and records exist and cleaning is carried out with scheduled regularity or according to parameters such as slope, pipe age, etc.	T. 6	2
5	A service is available 24 hours a day to manage anomalies in the wastewater collection network.	T. 5	3
6	A protocol exists to deal with and resolve anomalies in the wastewater collection network.	T. 6	3
7	A record of anomalies in the wastewater collection "system" and the processes followed to resolve them exists (documented at least on paper).	T. 5	1
8	A computerized system for recording and managing anomalies in the wastewater collection "system" is available.	T. 3	2
9	A GIS covering the entire wastewater collection network and its elements exists, as do protocols to update it systematically.	T. 3	3

10	A system exists to record and manage anomalies in the GIS, which also shows users connected to the wastewater network.	T. 3	2

SQ3.3 Supervision and control of the wastewater collection service

Type: Best Practices
Service: Sanitation
Normalization: Weighted by practices
Glossary: Real time
Definition: Includes:

Practices	Reliability	Weight
1 Devices to measure flow rates in the wastewater collection network and collector mains are available (at least one device per 20,000 inhabitants).	T. 1	1
2 A "real-time" telemetry system is available to manage the wastewater collection network.	T. 3	1
3 A record of measurements taken and alarms produced in the wastewater collection network is kept.	T. 6	2
4 Flow-regulating elements (e.g. remotely controlled gates) exist in the collector mains or wastewater collection network.	T. 1	1
5 Systems exist to support decision-making in regular and exceptional operation of the drainage or wastewater collection network.	T. 3	1

SQ3.4 Time taken to resolve "incidents" in the wastewater collection network

Considers the average time taken to resolve an incident in any element or component of the sewerage network (connections, sewers and collectors) from the moment the incident is reported until it is definitively resolved.

It is assumed that any "incident" potentially affects individual wastewater collection service availability and that all incidents represent a disruption to proper operation of the "system" to be rated.

Only incidents considered fortuitous, or that have not been classified as attributable to third parties, will be considered.

Definition: Average time taken to resolve "incidents" considered fortuitous in the wastewater collection network in the full calendar year preceding the rating date.
Type: Indicator
Service: Sanitation
Glossary: System, Incident
Formula: [CS3-V1]   Unit: hours
Normalization Function:

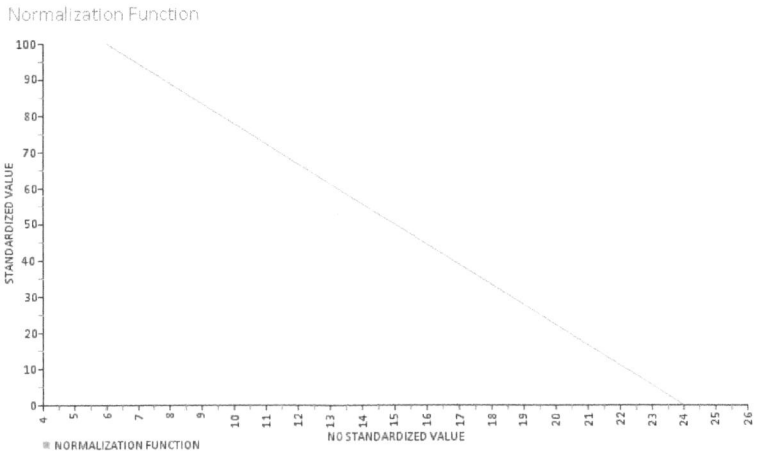

Variables

[**CS3-V1**] Average time taken to resolve "incidents" in the wastewater collection network
Definition: Average time taken to resolve "incidents" in the wastewater collection network considered fortuitous and occurring in the last full calendar year.
Units: hours
Reliability: Table 15

SQ3.5 Time taken to connect to the wastewater service

User connection to the wastewater collection network is assessed by the length of time it takes to complete connection, considering neither delays attributable to issue of permits required to implement the works needed nor delays due to other obstacles for which the utility is not responsible.

Definition: Average time elapsed between completing application for the service and "completion" of connection of the new user's property to the public wastewater network. Value calculated for the full calendar year preceding the rating date.
Type: Indicator
Service: Sanitation
Glossary: Finished works
Formula: [CS3-V2]   Unit: days
Normalization Function:

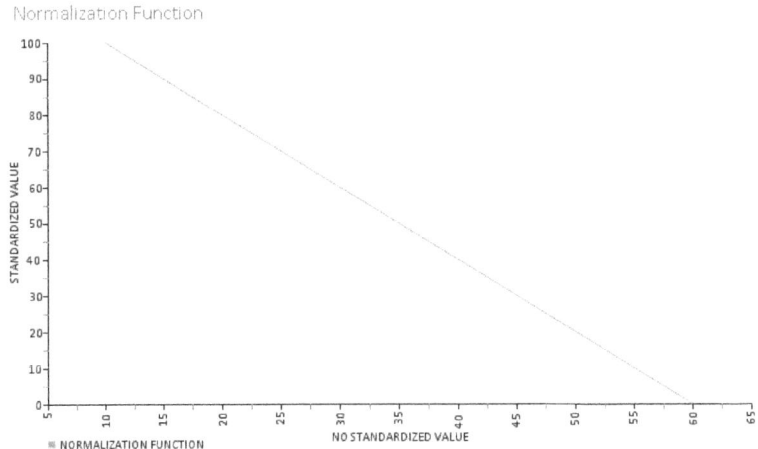

Variables

[CS3-V2] Average time elapsed between completing application for the service and "completion" of connection of the new user's property to the public wastewater network.

Definition: Average time elapsed between completing application for the service and "completion" and notification of connection of the new users' property to the public wastewater network (carried out in the last full calendar year). The time required to obtain works licenses and permits for which the utility is not exclusively responsible will be subtracted from the total.

Units: days

Reliability: Table 16

SQ3.6 Stormweather "incidents"

This assessment element evaluates operation of the stormwater drainage function of the wastewater collection system for which the utility is responsible, irrespective of whether this function is performed as an integrated or combined, direct or indirect or joint or separate part of the service.

Definition: Number of "malfunctions" in urban wastewater collection and drainage system elements that have substantially or visibly disrupted users, citizens, traffic or normal operation of public thoroughfares per 10,000 inhabitants in the "geographical area to be rated" for wastewater collection.

Although only malfunctions that would not have occurred had design or maintenance been appropriate should be considered, given the difficulty of identifying the causes in each case all recorded malfunctions will be considered, except those that coincide with rainfall intensities, recorded by an official body, above those established by design standards.

Type: Indicator
Service: Sanitation
Glossary: Incident, Malfunction, Geographical area to be rated
Formula: [CS3-V3]/[CS3-V4]*10.000   Unit: N. per 10.000 population
Normalization Function:

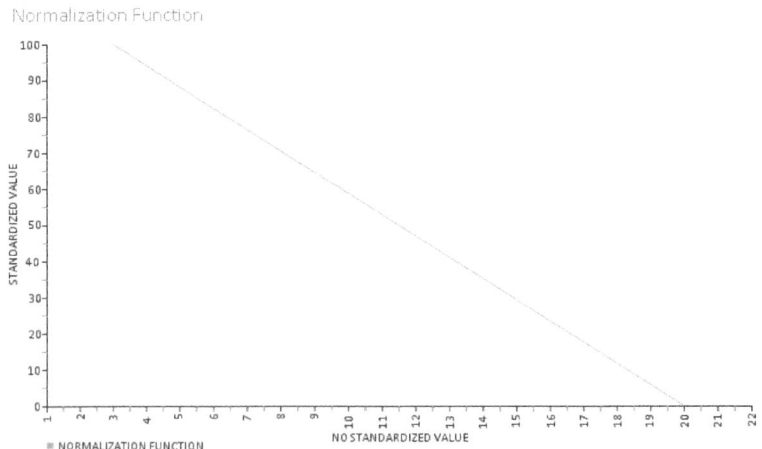

Variables

[CS3-V3] Number of "malfunctions" in urban wastewater collection and drainage system elements that have substantially or visibly disrupted users, citizens, traffic or normal operation of public thoroughfares (in the last full calendar year).

Definition: Number of "malfunctions" in urban wastewater collection and drainage system elements that have substantially or visibly disrupted users, citizens, traffic or normal operation of public thoroughfares (in the last full calendar year). All recorded cases will be considered, except those that coincide with rainfall intensities, as recorded by an official body, above those established in design specifications.

Units: -

Reliability: Table 122

[CS3-V4] Total inhabitants in the "geographical area to be rated" for wastewater collection.

Definition: Total inhabitants in the "geographical area to be rated" for wastewater collection at the end of the calendar year preceding the rating date.

Units: inhabitants

Reliability: Table 101

SQ4 User service

This section assesses user service quality, i.e. interaction between the service provider and the user as 'customer' of those services. It assesses forms of service; handling of complaints, requests, payments and other procedures; utility commitment to the service and user information about interruptions to the service and other contingencies. All these will have repercussions on the main objective of user satisfaction, which is also included in the assessment. The section considers both good practices and quantitative indicators to rate the aforementioned features.

Practices

SQ4.1   "Complaint" management and user satisfaction monitoring
SQ4.2   User service quality
SQ4.3   Commitment to user service and "contingency" information

Indicators

SQ4.4   Perception of general user satisfaction
SQ4.5   User perceptions of problem resolution quality
SQ4.6   Number of "customer service complaints" per 100 users and year
SQ4.7   Customer call service waiting time
SQ4.8   Customer service center waiting time
SQ4.9   Time taken to resolve problems

SQ4.1 "Complaint" management and user satisfaction monitoring

Type: Best Practices
Service: Drinking Water and/or Sanitation
Normalization: Weighted by practices
Glossary: Complaint, Technical expertise
Definition: Includes:

Practices	Reliability	Weight
1	An integrated complaint management system is in place which besides recording "complaints" (by any means) also tracks complaint resolution.	T. 3	1
2	Resolution of all "complaints" is notified and satisfaction is verified.	T. 6	1
3	"Complaint" and resolution records are analyzed at least quarterly and results of this analysis are used to improve service and user management.	T. 6	3
4	A user satisfaction survey is conducted once a year among utility service users. It complies with the following characteristics: (i) The survey is statistically representative of the user population; (ii) The survey methodology is stable and reproducible; (iii) The survey is conducted by a group/organization with appropriate "technical expertise"; (iv) The survey is conducted by a third party.	T. 2	2
5	Satisfaction with resolution of user "complaints" is monitored constantly (representative sample of all users making complaints).	T. 6	1

SQ4.2 User service quality

Type: Best Practices
Service: Drinking Water and/or Sanitation
Normalization: Weighted by practices
Glossary: Complaint
Definition: Includes:

	Practices	Reliability	Weight
1	A call center exists to receive customer and technical "complaints" and process complaints and contract-related issues. The call center operates during workdays and working hours.	T. 5	1
2	The call center operates 24/7.	T. 5	1
3	The call center has a sufficient number of duly trained staff provided with computer support.	T. 5	1
4	The rate charged for calls to the call center does not surpass the local call rate.	T. 5	1
5	Customer service centers have a sufficient number of duly trained staff provided with computer support.	T. 5	1
6	The website is operative and at least 4 of the following procedures can be performed online: (i) Check account/bill; (ii) Make payment online; (iii) Submit a "complaint"; (iv) Establish a new contract; (v) Request a service feasibility check; (vi) Request service stoppage/termination.	T. 5	1
7	The utility admits at least 3 of the following payment channels: (i) Internet (institutional website and others); (ii) Automatic payment charged to bank account or credit card; (iii) Telephone; (iv) Payment centers (at utility's offices or other defined locations).	T. 5	1

AquaRating

SQ4.3 Commitment to user service and "contingency" information

Type: Best Practices
Service: Drinking Water and/or Sanitation
Normalization: Weighted by practices
Glossary: Contingency, Complaint
Definition: Includes:

Practices	Reliability	Weight
1 A customer/user ombudsman exists.	T. 2	2
2 A duly publicized commitment charter exists (e.g. published on the website, sent to users, etc.) establishing the utility's commitment to respond to "complaints" within the timeframe established by the regulator in applicable cases.	T. 2	2
3 The utility commits to compensate users in the event of non-compliance with the terms established in the commitment charter.	T. 2	1
4 Users are notified at least 48 hours in advance of programmed service stoppages and planned interruptions, receiving notification at their homes or buildings or by any other means of communication that ensures that users receive notification. For wide-area notification, the utility may use radio and television broadcasts. A complete database for managing communication with users exists and procedures are in place to ensure it is up to date.	T. 2	1
5 Users are informed about progress and expected resolution of unforeseen "contingencies" (e.g. by means of automated calls or text messages to users' phones, automated responses from the utility's user service line, website, etc.).	T. 2	1
6 Critical users (hospitals, schools, high-volume consumers, etc.) are identified and a special procedure is in place to provide them with timely information about interruptions, expected service impact and progress and expected resolution of unforeseen "contingencies".	T. 2	2

SQ4.4 Perception of general user satisfaction

Definition: Percentage of "users satisfied" with the service in general.
Type: Indicator
Service: Drinking Water and/or Sanitation
Glossary: Satisfied user
Formula: ([CS4-V1]/[CS4-V2])*100   Unit: %
Normalization Function:

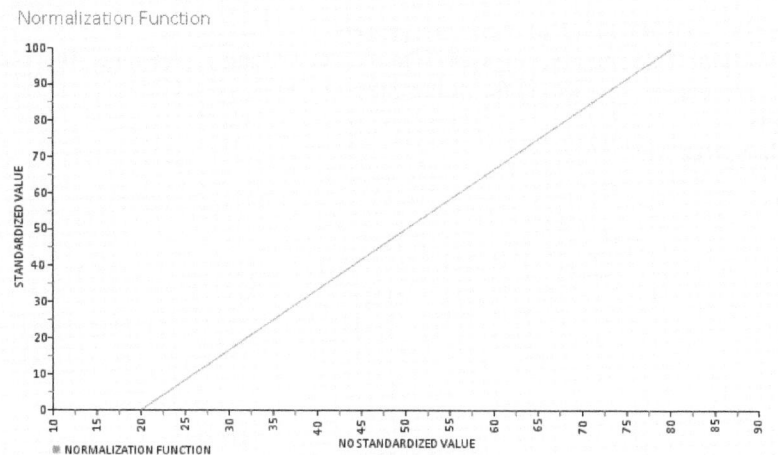

Variables

[CS4-V1] Number of users among those interviewed "satisfied" with the service in general
Definition: Number of users among those interviewed "satisfied" with the service in general (taking as reference the last full calendar year preceding the rating date).
Units: -
Reliability: Table 18

[CS4-V2] Total number of users interviewed
Definition: Total number of users interviewed (taking as reference the last full calendar year preceding the rating date).
Units: -
Reliability: Table 56

SQ4.5 User perceptions of problem resolution quality

Definition: Percentage of users who have experienced a problem and express "satisfaction" with the quality of the solution provided.
Type: Indicator
Service: Drinking Water and/or Sanitation
Glossary: Satisfied user
Formula: ([CS4-V3]/[CS4-V4])*100   Unit: %
Normalization Function:

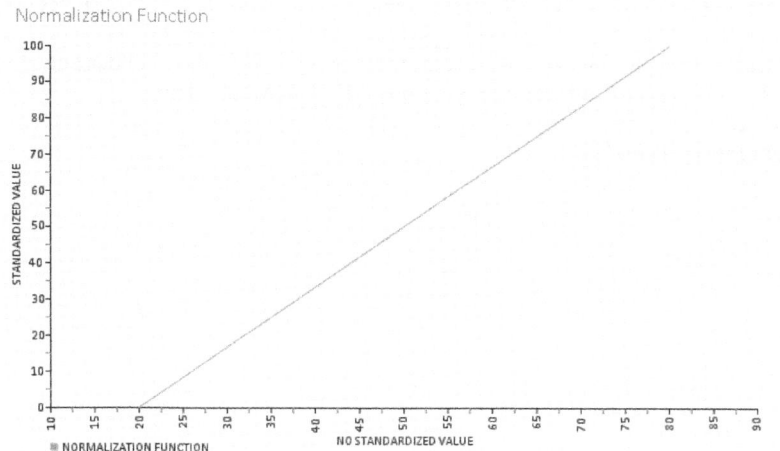

Variables

[CS4-V3] Number of users interviewed who have experienced a problem and express "satisfaction" with the quality of the solution provided.
Definition: Number of users interviewed who have experienced a problem and express "satisfaction" with the quality of the solution provided (taking as reference the last full calendar year preceding the rating date).
Units: -
Reliability: Table 19

[CS4-V4] Total number of users interviewed who have experienced a problem.
Definition: Total number of users interviewed who have experienced a problem in the last full calendar year preceding the rating date.
Units: -
Reliability: Table 56

SQ4.6 Number of "customer service complaints" per 100 users and year

Definition: Annual number of "customer service complaints" / Number of "registered users" multiplied by 100 in the calendar year preceding the rating date.
Type: Indicator
Service: Drinking Water and/or Sanitation
Glossary: Registered user, Customer service complaint
Formula: ([CS4-V5]/[CS4-V9])*100   Unit: %
Normalization Function:

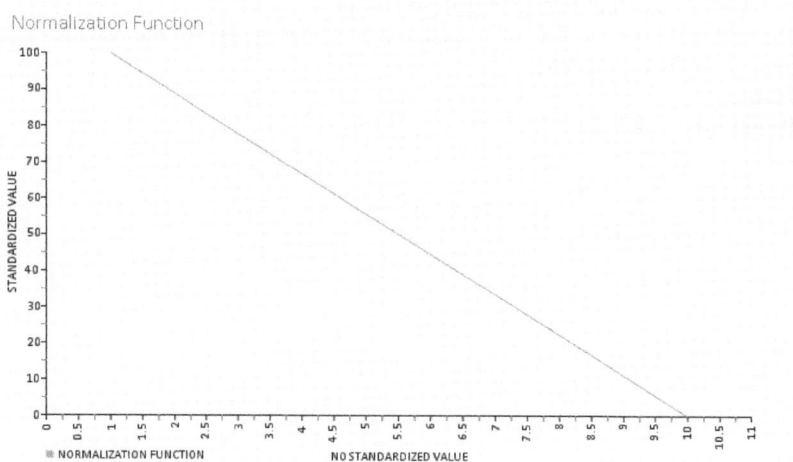

Variables

[CS4-V5] Number of "customer service complaints"
Definition: Number of "customer service complaints" in the calendar year preceding the rating date related to aspects not linked to proper operation of water supply (in terms of appropriate water quantity or quality and wastewater collection and treatment). In case of doubt, complaints will be considered customer service complaints and added to this variable.
Units: -
Reliability: Table 20

[CS4-V9] Total number of "registered users"
Definition: Total number of "registered users" at the end of the calendar year preceding the rating date.
Units: -
Reliability: Table 21

SQ4.7 Customer call service waiting time

Definition: Average user waiting time during calls to the call center (data for the last full calendar year preceding the rating date).
Type: Indicator
Service: Drinking Water and/or Sanitation
Glossary:
Formula: [CS4-V6]   Unit: minutes
Normalization Function:

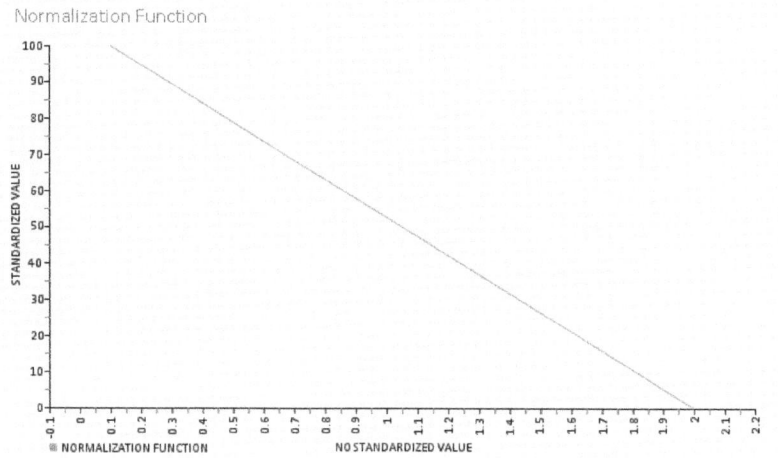

Variables

[CS4-V6] Average user waiting time during calls to the call center
Definition: Average user waiting time during calls to the call center (data for the last full calendar year preceding the rating date).
Units: minutes
Reliability: Table 22

SQ4.8 Customer service center waiting time

Definition: Average user waiting time when visiting the customer service center (for any reason except payments).
Type: Indicator
Service: Drinking Water and/or Sanitation
Glossary:
Formula: [CS4-V7]   Unit: minutes
Normalization Function:

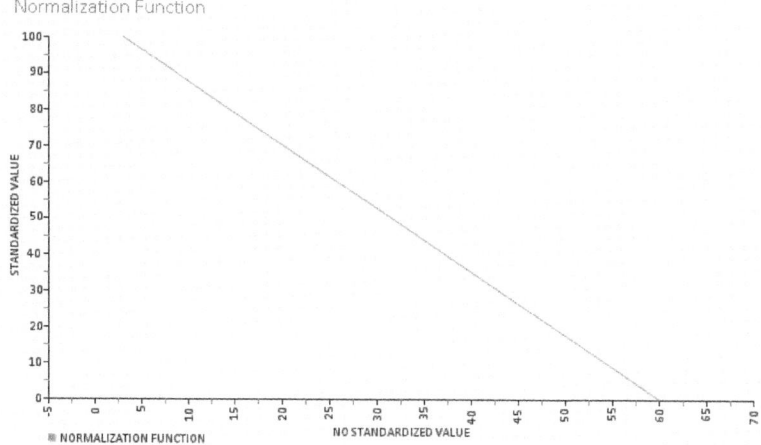

Variables

[CS4-V7] Average user waiting time when visiting the customer service center

Definition: Average user waiting time when visiting the customer service center. All reasons for visiting are considered, except those regarding payment (data for the last full calendar year preceding the rating date).

Units: minutes

Reliability: Table 23

SQ4.9 Time taken to resolve problems

Definition: Average time elapsed between submission of a "customer service complaint" and resolution of the problem.
Type: Indicator
Service: Drinking Water and/or Sanitation
Glossary: Customer service complaint
Formula: [CS4-V8]   Unit: working days
Normalization Function:

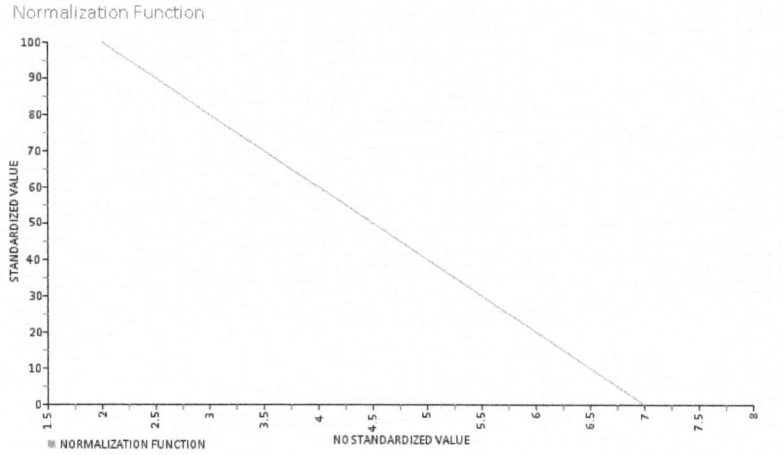

Normalization Function

Variables

[CS4-V8] Average time elapsed between submission of a "customer service complaint" and resolution of the problem.
Definition: Average time elapsed between submission of a "customer service complaint" and resolution of the problem (data for the last full calendar year preceding the rating date).
Units: working days
Reliability: Table 24

PE Investment Planning and Implementation Efficiency

One of utilities' main duties is to provide water and wastewater services throughout the geographical service coverage area. For this purpose, they have to plan considerable investment to augment or sustain "system" capacity and to extend or improve drinking water supply and distribution "systems" and wastewater collection and treatment. The purpose of this area is to rate the utility insofar as it has adequate investment plans in place for each "system" for which it is responsible and to assess implementation efficiency. Only with adequate planning will the utility be able to meet its short-, medium- and long-term needs and achieve the service quality and coverage goals set by the authority.

In several countries, having an investment plan in place is a requirement established by the regulator as it forms the basis for tariff setting as well as for monitoring the utility's compliance with its obligations.

Irrespective of the circumstances, these plans are indicative of the utility's future actions and are not a tool that constrains decision-making. Therefore, plan formulation and implementation must include the adjustments and flexibility needed to deal with unforeseen requirements that arise over time. Nevertheless, having a good investment plan in place is considered to be critical for good business decision-making, particularly in this field, which requires considerable investment in long-term infrastructure.

Assessment considers five fundamental criteria: existence of a documented investment plan with minimum content; appropriateness of the methodology used to develop it; the way in which it will be implemented over time; special inclusion of physical asset management; contingency planning for emergencies due to natural or other disasters; and existence of research and development.

The sub-areas of evaluation are:

PE1 Investment plan content and efficiency
PE2 Investment plan implementation efficiency
PE3 Existing physical asset management efficiency
PE4 Emergency planning
PE5 Research and development

PE1 Investment plan content and efficiency

Water and wastewater services face major challenges in delivering the quantity and quality demanded by users, which in turn creates a need to plan major investment. Therefore, it is essential to rate the utility regarding the existence of an investment plan, which should be duly formulated and approved by the utility's higher authorities and fully monitorable. Additionally, this plan must seek to incorporate all appropriate technical, financial, environmental and social methodological elements to ensure that the general condition of providing the best "system" expansion solution is met at least possible cost. The PE1 sub-area rates the content of the plan itself, as well as its quality, by rating the methodological appropriateness of each formulation stage (diagnosis, identification of alternatives, choice of solutions, financial analysis), utilizing for this purpose project formulation and assessment models well known to the sector.

Practices

PE1.1 Investment plan contents
PE1.2 Diagnosis methodology
PE1.3 Methodology for identifying and analyzing alternatives and defining solutions
PE1.4 Methodology for analyzing the plan's financial aspects

PE1.1 Investment plan contents

Type: Best Practices
Service: Drinking Water and/or Sanitation
Normalization: Weighted by practices
Glossary: System, Board of directors, Service stage, Works (in relation to investment plan projects)
Definition: The investment plan should be documented and contain the following elements either in the main text, appendices or complementary documents:

	Practices	Reliability	Weight
1	The investment plan defines general and specific objectives.	T. 25	1
2	The investment plan establishes goals for coverage and quality for each service and "system" in the utility's rating area.	T. 25	1
3	The investment plan establishes the budget for the "works" detailed in the plan, including planned sources of finance.	T. 25	1
4	The investment plan includes detailed definitions of the programs (sets of projects) or specific projects to be developed.	T. 25	1
5	The plan contains an implementation schedule.	T. 25	1
6	The plan shows the general configuration of existing and planned "systems", by "service stage", on figures or maps at adequate scale.	T. 26	1
7	The plan includes a general description of the criteria used in its formulation.	T. 25	1
8	The plan contains a list of programs or projects, per "system", specifically indicating: name, estimated cost, year of operation start-up and observations or comments.	T. 25	1
9	The plan includes a summary table of the investment program indicating for every "service stage" and year the project and annual investment.	T. 25	1

10	The plan as a whole is approved by the "board of directors" or competent authority (updates for individual works may be approved separately).	T. 27	3

PE1.2 Diagnosis methodology

Type: Best Practices
Service: Drinking Water and/or Sanitation
Normalization: Weighted by practices
Glossary: System, Supply-demand balance, Geographical area to be rated
Definition: Plan formulation methodology is presented in the plan, appendices or complementary documents for each "system" in the "geographical area to be rated". The contents of this documentation include:

	Practices	Reliability	Weight
1	Up-to-date digital maps (updated less than 1 year before the plan's approval date) exist identifying the "geographical area to be rated". These maps have an adequate scale (minimum 1:10,000) and show UTM coordinates, urban boundaries, main streets or avenues, contour lines, ground elevation and noteworthy geographical features. These maps include the location of relevant infrastructure works affecting the drinking water and wastewater "systems".	T. 56	1
2	Projections exist of population size, user numbers and demand in the short, medium and long term (approx. 5, 10 and 15 years, respectively).	T. 25	1
3	Current and expected population size and user numbers in the "geographical area to be rated" are documented. These are segmented by user type and current and expected location according to overall and territorial growth projections for the service area, including, if applicable, real estate projects.	T. 25	1
4	Coverage and service quality goals exist. These are equal to or higher than those set by the authority (governing entity and/or regulatory entity), if applicable. If individual solutions are considered, the number of households served per solution type is specified.	T. 25	1
5	Scenarios establishing the range of projected demand variation exist, considering, among other aspects, support programs for low-income families to connect them to the sewerage system, if applicable.	T. 25	1
	Consumption histories (daily per capita water consumption		

6	rate) exist by user type and consumption projections according to changes in income levels, water use technologies, demand management programs and other relevant factors are documented.	T. 25	1
7	Daily and hourly consumption coefficients, average annual and maximum volumes, infiltration rates and others are documented and supported by empirical information based on specific studies.	T. 25	1
8	A detailed water balance as per the IWA classification or similar exists (with differentiation between apparent and real losses and authorized and unauthorized consumption).	T. 25	1
9	Information about drinking water in each "system" is summarized in tables that show the following for every year up to the analysis horizon (annual data may be calculated by interpolating 5-year estimates or others): total population; user population (population served); number of drinking water users; annual consumption volumes; average consumption flow rates (peak daily flow and peak hourly flow); volume of losses; average production flow rates (peak daily flow and peak hourly flow).	T. 25	1
10	Information for each wastewater "system" is summarized in tables that show the following for every year up to the analysis horizon: total population; user population (population served); number of users; daily per capita water consumption rate; recovered flow; infiltrated flow, average flow; peak hourly flow; projected load (KgDBO5/day).	T. 25	1
11	A "supply-demand balance" is calculated for each component of the "system" to establish the years in which deficits occur in order to consider actions or works to eliminate or reduce the projected deficits.	T. 25	3
12	The analysis considers the condition of existing infrastructure and determines whether it needs replacing.	T. 25	3

PE1.3 Methodology for identifying and analyzing alternatives and defining solutions

Type: Best Practices
Service: Drinking Water and/or Sanitation
Normalization: Weighted by practices
Glossary: Applicable regulations, System, Service stage, In force (investment plan in force)
Definition: Corresponds to documents which contain the essential criteria used to identify and define solutions, such as appendices or complementary documents to the investment plan "in force". These documents include:

	Practices	Reliability	Weight
1	Solutions required by the "system" to satisfy proposed coverage and quality goals are defined. Solutions are developed at a prefeasibility level (cost confidence level ±15%) and take into consideration analysis of supply-demand for each component, condition of existing infrastructure, regulations in force and other elements. These solutions are analyzed and presented in documents approved by the competent unit.	T. 25	1
2	Possible solutions are subject to multi-criteria assessment of alternatives that explicitly analyzes the available options, taking into consideration definition and assessment of the available project options in terms of configuration, sizing, analysis horizon, technological options, environmental or other restrictions, service delivery and environmental regulations, optimal date of operation start-up, restrictions due to preparation and implementation times, etc.).	T. 25	1
3	Possible solutions are subject to assessment of alternatives that applies algorithms to minimize investment costs and incremental operating and maintenance costs.	T. 25	1
4	Possible solutions are subject to assessment of alternatives that applies facility and service safety, risk and vulnerability criteria.	T. 25	1
	Possible solutions are subject to assessment of alternatives that takes into consideration life cycle assessment (LCA), CO_2 emissions, environmental sustainability and mitigation		

5	of other externalities.	T. 25	1
6	Possible solutions are subject to assessment of alternatives that uses duly calibrated and updated tools to analyze hydraulic or hydrological behavior and employs geo-referenced database systems.	T. 25	1
7	In sections that concern drinking water supply, analysis of current and future water sources is carried out to determine available flows, quality and water rights.	T. 25	1
8	In sections that concern drinking water supply, specific proposals to reduce unaccounted for water are included.	T. 25	2
9	In sections that concern sanitation, analysis of wastewater is carried out to determine treatment level and impact of discharges on the receiving body.	T. 25	1
10	In sections that concern sanitation, treatment options are analyzed in terms of "applicable environmental regulations".	T. 25	2
11	In the overall configuration of the plan, the solution considered the best alternative is the one with the lowest cost at present value that complies with all restrictions concerning demand, environment and other applicable requirements. It specifically includes explicit presentation of the selected alternative and the corresponding minimum overall cost analysis.	T. 25	2
12	Information about selected projects is presented in a detailed list by "service stage".	T. 25	1
13	Information about selected projects is presented in a detailed list with corresponding costs.	T. 25	1
14	Information about selected projects includes operation start-up deadlines.	T. 25	1

PE1.4 Methodology for analyzing the plan's financial aspects

Type: Best Practices
Service: Drinking Water and/or Sanitation
Normalization: Weighted by practices
Glossary: Discount rate applicable to capital cost
Definition: The methodology for considering financial aspects is included in the plan, appendices or complementary documents and considers the following:

	Practices	Reliability	Weight
1	A financial assessment is carried out for each project or set of projects, calculating the net present value (NPV), which includes a projection for each year in the investment analysis horizon, investment in replacement, incremental operating and maintenance costs, and revenue from service delivery based on present tariffs (or expected, if applicable). The "discount rate applicable to the utility's capital cost" is used.	T. 25	1
2	A financial assessment is carried out for the plan as a whole, determining the NPV and including any financing restrictions the utility may have.	T. 25	1
3	The profitability analysis of both the individual projects and the set (irrespective of whether the outcome is negative or positive) informs project selection and plan formulation.	T. 25	1
4	The final plan solely considers projects with a positive NPV, and the plan itself has a positive NPV; or the overall plan has a positive NPV, even though some projects are not profitable. This assessment is approved by the competent unit.	T. 25	1

PE2 Investment plan implementation efficiency

Efficient implementation of plans, projects and works deserves special attention in the rating system, since it reflects material use of investment resources. This aspect of water and wastewater utility management is considered in sub-area PE2.

To assess this sub-area three good practice assessment elements and one indicator are used. The first one identifies and rates practices used in tracking projects, their costs and their implementation periods. The second measures compliance with the investment plan produced in year 0 (considered the reference point). Naturally, this does not prevent continual updating of the investment plan. However, this indicator seeks to ascertain the quality of the planning proposed under the previous criterion. The next element assesses the degree of deviation between the final cost of the works implemented and the initial cost tendered. Cost overruns during the most critical phase of works, namely implementation, are assessed via this measurement, which also captures final design preparation quality. The last element assesses deviations between works' real implementation time and tendered times, since delays in implementation have a negative impact on various aspects of utility operation.

Practices

PE2.1 Systems for monitoring implementation of investment plan projects

PE2.3 Degree of cost variation in "completed works"

PE2.4 Degree of deviation from deadlines established for implementation of "works"

Indicators

PE2.2 Compliance with the investment plan

PE2.1 Systems for monitoring implementation of investment plan projects

Type: Best Practices
Service: Drinking Water and/or Sanitation
Normalization: Weighted by practices
Glossary: Works (in relation to investment plan projects), Integrated project monitoring system
Definition: This assessment element includes practices to verify and monitor aggregate costs and timeframes at the end of the full calendar year preceding the rating date for all projects proposed in the investment plan in year 0. In particular, "Integrated monitoring systems" exist to ensure the following:

Practices	Reliability	Weight
1	An "integrated monitoring system" exists that monitors investment plan projects formulated in year 0 and projects or "works" resulting from it.	T. 3	1
2	An "integrated monitoring system" exists that records costs of planned projects (established at prefeasibility level) and costs of projects (or works) at implementation design (or final design) level.	T. 3	1
3	An "integrated monitoring system" exists that records costs of planned projects (or works) at tender level in relation to costs of projects (or works) at implementation design (or final design) level. Causes for discrepancies over ±20% are recorded.	T. 3	1
4	An "integrated monitoring system" exists that records final costs of projects (or works) in relation to tendered costs. Causes for discrepancies over ± 20% are recorded. Final costs include all types of modification to the tendered project.	T. 3	3
5	An "integrated monitoring system" exists that monitors deadlines set for start-up of investment plan projects planned for year 0 and real dates. Causes for discrepancies in excess of 6 months are recorded.	T. 3	3

Information on cost and deadline discrepancies for projects or works is reported to senior management at least once a

| 6 | year. | T. 2 | 3 |
| 7 | Systematic information feedback exists between the project cycle and the project planning and preparation process, allowing for adoption of corrective measures in order to reduce discrepancies in relevant indicators. This includes analysis of discrepancies in estimated demand, unit costs, etc. | T. 6 | 2 |

PE2.2 Compliance with the investment plan

Definition: Percentage representing expenditure disbursed for investment plan projects, either under implementation or completed, from year 0 of the plan until the end of the calendar year preceding the rating date as a proportion of projected investment plan expenditure for the same period. Only projects identified in the investment plan are considered.
Type: Indicator
Service: Drinking Water and/or Sanitation
Glossary:
Formula: ([EP2-V1]/[EP2-V2])*100   Unit: %
Normalization Function:

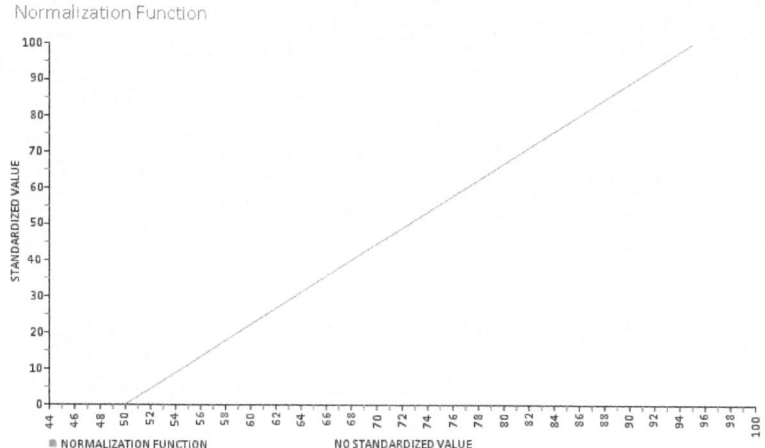

Variables

[EP2-V1] Expenditure disbursed for investment plan projects, either under implementation or completed

Definition: Expenditure disbursed for investment plan projects, either under implementation or completed, from year 0 of the plan until the end of the calendar year preceding the rating date.

Units: local currency

Reliability: Table 29

[EP2-V2] Projected expenditure on investment plan

Definition: Projected expenditure on the investment plan, from year 0 of the plan until the end of the calendar year preceding the rating date.

Units: local currency

Reliability: Table 30

PE2.3 Degree of cost variation in "completed works"

Assesses cost deviations for completed works in relation to the tendered budget. Cost comparison is made in real terms (adjusted for general inflation).

Investment plan projects are usually subdivided into works, which are put out to public tender. Several projects may be grouped into a single works contract for tendering purposes. This is why this indicator refers to the costs of the works and not necessarily to the projects in the investment plan.

Type: Best Practices
Service: Drinking Water and/or Sanitation
Normalization: Weighted by practices
Glossary: Finished works
Definition: Percentage of "works completed" in the calendar year preceding the rating date whose costs differ significantly from the budgeted value in the contract awarded (in the case that no works were completed in the calendar year preceding the rating date, the last calendar year in which works were completed is used). Each of the practices complied with (degree of cost variation) should be indicated.

The calculation requires the existence of a list of the works completed during the analysis period, indicating for each of them the cost at the time of award and the final cost, deducting adjustments for inflation. Based on this, works are distributed according to the percentage difference between final costs and award costs.

Practices		Reliability	Weight
1	Less than 50% of contracts produce differences between "final" cost and cost at award of more than ±30%.	T. 31	2
2	Between 60% and 70% of contracts produce cost differences between "final" cost and cost at award of up to ±25%.	T. 31	2
3	Between 80% and 90% of contracts produce cost differences between "final" cost and cost at award of up to ±20%.	T. 31	3
4	More than 90% of contracts produce cost differences between "final" cost and cost at award of up to ±10%.	T. 31	3

PE2.4 Degree of deviation from deadlines established for implementation of "works"

Type: Best Practices
Service: Drinking Water and/or Sanitation
Normalization: Weighted by practices
Glossary: Works (in relation to investment plan projects), Finished works
Definition: Percentage of "works completed" in the calendar year preceding the rating date whose implementation time differs significantly from the deadline established in the tender. This measures deviations in implementation timeframes in relation to the timeframes established in the tender (in the case that no works were completed in the calendar year preceding the rating date, the last calendar year in which works were completed is used). Each of the practices complied with (degree of deviation from deadlines) should be indicated.

The calculation requires the existence of a list of the works completed during the analysis period, indicating for each of them the real implementation time and the deadline established in the tender. Based on this, works are distributed according to the percentage difference between real timeframes and tender deadlines.

Practices		Reliability	Weight
1	Less than 50% of contracts produce differences between actual implementation times and tendered implementation times of more than 24 months.	T. 32	2
2	Between 60% and 70% of contracts produce differences between actual implementation times and tendered implementation times of up to 18 months.	T. 32	2
3	Between 80% and 90% of contracts produce differences between actual implementation times and tendered implementation times up to 12 months.	T. 32	3
4	More than 90% of contracts produce differences between actual implementation times and tendered implementation times of up to 6 months.	T. 32	3

PE3 Existing physical asset management efficiency

Water and wastewater utilities possess an extremely significant stock of high-value infrastructure that must be managed appropriately to preserve and increase its value. For this reason, it is necessary to assess management of fixed physical assets. This assessment is achieved first by considering the practices followed by the operator in managing its assets. This measures the appropriateness of management actions and their sustainability over time (optimal asset management naturally considers planning and implementation of preventive maintenance. These actions are assessed under the Operating Efficiency area). A second element (indicator) measures the annual expenditure incurred by the operator to replace existing assets in order to preserve their value.

Practices

PE3.1   Physical asset management

Indicators

PE3.2   Annual investment in replacement of fixed physical assets

PE3.1 Physical asset management

Type: Best Practices
Service: Drinking Water and/or Sanitation
Normalization: Weighted by practices
Glossary:
Definition: Includes:

	Practices	Reliability	Weight
1	Records exist of existing infrastructure and its condition: survey of fixed assets' operational capacities and condition (good, fair or poor).	T. 33	3
2	A plan exists for maintenance and replacement of fixed physical assets based on failure risk analysis, costs, etc.	T. 2	5
3	Up-to-date handbooks exist, and are used, detailing operation and maintenance of fixed physical assets.	T. 6	1
4	Corresponding staff are trained to manage the fixed physical assets.	T. 4	1
5	Fixed physical asset management is specifically reflected in the utility's strategic plan (see ME1) and a unit is assigned responsibility for it.	T. 34	1

PE3.2 Annual investment in replacement of fixed physical assets

Definition: Percentage representing annual investment in replacement of fixed physical assets as a proportion of the gross value of fixed physical assets at the beginning of the year (average for the last 3 full calendar years).
Type: Indicator
Service: Drinking Water and/or Sanitation
Glossary: Geographical area to be rated
Formula: ([EP3-V1]/[EP3-V2])*100   Unit: %
Normalization Function:

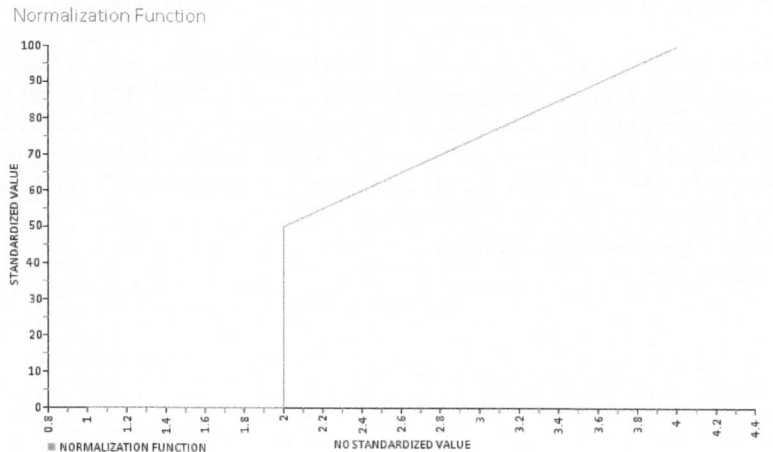

Variables

[EP3-V1] Annual investment in replacement of fixed physical assets
Definition: Cost of substituting or replacing relevant components or parts of the facilities, equipment and infrastructure used for service provision in the "geographical area to be rated", including the cost of substituting or replacing infrastructure not owned by the utility in the case that the utility is responsible for such replacement costs.
Units: financial statement currency
Reliability: Table 35

[EP3-V2] Gross value of total fixed physical assets at the beginning of the year
Definition: Gross value of the facilities, equipment and infrastructure linked to service provision in the "geographical area to be rated" (except land), including infrastructure not owned by the utility, in the case that the utility is responsible for its replacement and maintenance costs. The gross value must match the gross value entered in the accounts at the beginning of the year, including value adjustments if applicable.
Units: financial statement currency
Reliability: Table 36

PE4 Emergency planning

Availability of appropriate drinking water and wastewater services is an essential social need. Therefore, ensuring operation in case of natural disasters or other events such as drought, act of terrorism or similar, must be anticipated as part of infrastructure design and operation and appropriate plans must be formulated to deal with emergency situations. Criterion PE4 rates the quality of the plans and preparations the institution has in place in anticipation of this type of phenomenon, considering for this the availability and quality of a plan to deal with emergencies.

Practices

PE4.1   "Emergency" plan

PE4.1 "Emergency" plan

Type: Best Practices
Service: Drinking Water and/or Sanitation
Normalization: Weighted by practices
Glossary: System, Emergency
Definition: Plan containing the elements needed to deal with "emergencies".

	Practices	Reliability	Weight
1	An analysis exists detailing the main risks the utility faces, including the corresponding probability of occurrence.	T. 37	1
2	A vulnerability analysis exists specifying the "system" elements potentially most affected.	T. 37	1
3	Mitigation measures to reduce "system" vulnerability have been identified and implemented and are incorporated in the investment plan.	T. 38	1
4	An "emergency" plan exists identifying the entities responsible for issuing alerts and staff have been assigned to monitor them.	T. 37	1
5	An "emergency" plan exists that includes an incident command group and prior coordination agreements with other entities, and identifies users assigned high priority for service restoration.	T. 37	1
6	The "emergency" plan is comprehensively updated in the wake of the last event that affected the utility, following modifications of conditioning factors, or is validated with pre-established regularity.	T. 37	1
7	The "emergency" plan has been widely communicated to staff, who are trained to implement it.	T. 4	1

PE5 Research and development

Technological progress and the need to ensure quality service delivery at lower costs demand that modern utilities are duly aware of advances in the sector in this field and that they specifically allocate resources to research and development, especially to areas focusing on incorporating new processes or improving existing ones.

Practices

PE5.1   Research and development

Indicators

PE5.2   Investment in research and development

PE5.1 Research and development

Type: Best Practices
Service: Drinking Water and/or Sanitation
Normalization: Weighted by practices
Glossary:
Definition: Includes:

	Practices	Reliability	Weight
1	An R&D plan exists that contains: general and specific objectives, lines of work, goals, activity plan, budget and staff responsible for it.	T. 2	5
2	R&D activities include systematic collection and processing of information in order to improve internal practices.	T. 2	1
3	R&D activities include some form of proprietary research or contribution to it.	T. 2	1
4	A system exists that provides access to documents and publications and includes subscriptions to the main trade magazines and online information services.	T. 2	1
5	At least one person is in charge of gathering and communicating relevant news about R&D in the sector.	T. 2	1
6	A formal continuing professional development program exists for professionals and technicians: courses, conferences, specialist postgraduate studies, etc. (Note: this is a specific subset of general training).	T. 2	3
7	Specific actions are taken to share experiences with other utilities by means of study visits, seminars etc.	T. 2	1
8	Research and development agreements exist with local or international entities.	T. 2	1
9	An R&D activities monitoring system exists.	T. 2	2

PE5.2 Investment in research and development

Definition: Measures expenditure on (and/or investment in) R&D over a 3-year period as a percentage of the utility's total operating revenue in the same period.
Type: Indicator
Service: Drinking Water and/or Sanitation
Glossary:
Formula: ([EP5-V1]/[SF3-V12])*100   Unit: %
Normalization Function:

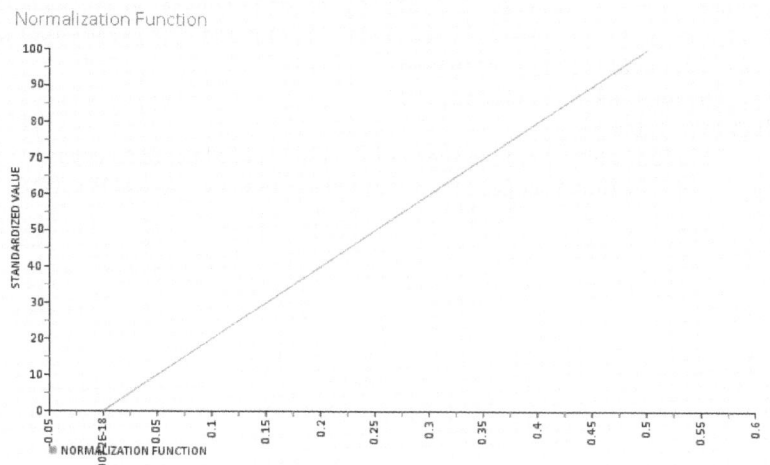

Variables

[EP5-V1] Expenditure on (or investment in) research
Definition: Expenditure on (or investment in) research which a utility decides to carry out with the intention of discovering what might lead to development of new products or processes, or to improvement of existing products and processes.
Units: financial statement currency
Reliability: Table 39

[SF3-V12] Revenue from service delivery
Definition: Revenue from service delivery or sales entered in the income statement or profit and loss statement for the period and consistent with the revenue entered in accounts receivable.
Units: financial statement currency
Reliability: Table 88
Note: For utilities for whom the scope of the Financial Statements does not coincide with the rating scope, a modified reliability table applies. See Table 1088

OE Operating Efficiency

Efficient use of the resources available to the utility is an essential part of delivering a high-quality public service. Efficient use of water and energy resources and efficient management of infrastructure and of operation and maintenance costs are the cornerstones of the Operating Efficiency area.

In particular, this area assesses efficiency in use of the various resources needed to comply with the service quality standards or objectives established in the regulatory framework or defined in the utility's strategic plan. However, compliance with service standards as such is assessed separately in the Service Quality area.

Assessment is geared towards measurement of use of the main resources (water, energy, others) employed in operation of available infrastructure, which will be the result of the infrastructure planning and management practices considered in other rating areas.

The sub-areas of evaluation are:

OE1 Water resource management efficiency
OE2 Energy usage efficiency
OE3 Infrastructure management efficiency
OE4 Operational and maintenance cost efficiency

OE1 Water resource management efficiency

Water resource management efficiency measures the degree of physical use of water resources introduced into the supply and distribution "system" (taken in from a natural source or imported from other "systems"). Used resources are understood to be those for which the final destination is a known and authorized point of consumption. Tank overflows, real losses, operational uses and frauds are not considered used water.

Assessment elements based on some frequently used parameters employed in efficiency management, such as apparent losses or unbilled water (non-revenue water) have not been included in this sub-area's assessment, since they are directly linked to other indicators of infrastructure efficiency or processes already considered in the appropriate areas.

Efficient use of water by private users on their own property and its relation to demand management is usually included in operating efficiency policies. However, in this rating system these issues are included in the environmental sustainability indicators. Its inclusion in this rating area would only make sense as a way of measuring compliance with possible demand management plan policies (voluntary or imposed by the regulator).

Practices

OE1.1   Control of water use and destinations
OE1.3   Management of real losses
OE1.5   Management of water used in operation
OE1.7   Management of "reclaimed water"

Indicators

OE1.2   Control of water at points of use and consumption
OE1.4   Real losses in the water supply, transportation and distribution infrastructure
OE1.6   Water used in operation
OE1.8   "Reused water"

OE1.1 Control of water use and destinations

Type: Best Practices
Service: Drinking Water
Normalization: Weighted by practices
Glossary: System, Property, Entry point into the drinking water supply system
Definition: Includes:

Practices		Reliability	Weight
1	Individual water flow rate or volume-metering devices (micro-metering) are installed at all points of use and consumption and these are read and the data recorded at least once a quarter.	T. 1	2
2	Water flow rate or volume-metering devices are installed at all "entry points to the water supply system" and these are read and the data recorded at least once an hour.	T. 1	3
3	Policies exist for dimensioning and renewing individual metering devices (micro-metering) that focus on maintaining the error levels or confidence intervals established in the regulations and on homogenizing metrological classes.	T. 6	1
4	Policies exist for dimensioning, renewing and verifying water flow rate or volume-metering devices at the "entry points to the system" that focus on maintaining the error levels or confidence intervals established in the regulations and that tend to homogenize classes, types and brands.	T. 6	1
5	Information on the location of all points of use and consumption is available in a geographical database of the distribution infrastructure.	T. 3	1
6	Distribution networks are sectored and inflow volumes to sectors are measured frequently (at least once an hour). Sector scope does not exceed 10,000 "properties". This practice is only considered to be complied with if more than 90% of network length is sectored.	T. 3	1

Water supply and controlled water consumption balances for the entire supply network are calculated and documented at

| 7 | least once a quarter. | T. 6 | 2 |

| 8 | Water supply and controlled consumption balances for all sectors are calculated and documented at least once a month, with consumption being calculated pro rata if consumption records are required for greater time intervals. | T. 6 | 1 |

| 9 | A procedure, unit or specific plan exists for reducing uncontrolled water. It includes, in addition to metering of all use and consumption, reduction of water use and consumption that does not generate revenue. | T. 6 | 1 |

| 10 | Reliability indicators exist for measurements of flow supplied to sectors and to the whole "system". | T. 6 | 1 |

OE1.2 Control of water at points of use and consumption

This indicator assesses the degree of control over the final destination of the water introduced into the supply "system" by measuring individual flow rates and volumes at each point of use and consumption. It reveals water balance reliability and geographical distribution. It reflects good practice and, in many cases, it is also the least disputable parameter among those usually employed to measure water resource usage efficiency. Similarly, it is the best starting point when diagnosing and implementing improvement measures. Individual consumption volumes that cannot be considered controlled water can only be assessed by means of estimates, which inevitably will be much less reliable than measured individual consumption volumes.

Uncontrolled water includes both real and apparent losses, fraud, operational uses and water used and consumed at a known point but not measured due to lack of an individual metering device, also known as authorized uses.

Definition: Percentage of "water introduced into the system" consumed and micro-metered as a proportion of total water introduced into the "system". Total water metered for individual consumption (whether users have a contract or not) in the full calendar year preceding the rating will be considered.
Type: Indicator
Service: Drinking Water
Glossary: System, Water volume incorporated into the system
Formula: ([EO1-V1]/[EO1-V2])*100   Unit: %
Normalization Function:

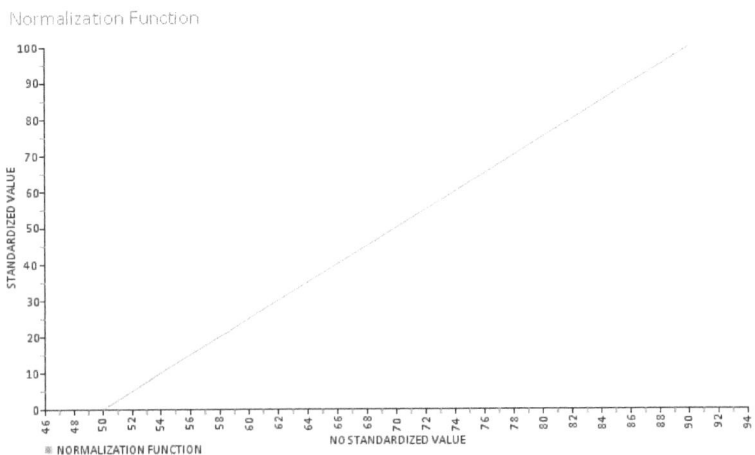

Variables

[EO1-V1] "Water volume introduced into the system" consumed at the points of use and consumption and micro-metered
Definition: "Water volume introduced into the system" consumed at the points of use and consumption and micro-metered
Units: m3
Reliability: Table 40

[EO1-V2] Total "water volume introduced into the system"
Definition: Total "water volume introduced into the system"
Units: m3
Reliability: Table 41

OE1.3 Management of real losses

Type: Best Practices
Service: Drinking Water
Normalization: Weighted by practices
Glossary: Incident
Definition: Includes:

Practices	Reliability	Weight
1 A utility unit is responsible for managing real losses or a well-defined operating procedure exists.	T. 2	1
2 A procedure exists for estimating real loss volumes based on standard criteria (IWA or similar, with differentiation between apparent and real losses and between authorized and unauthorized consumption) to estimate uncontrolled water components. Real loss volumes are computed at least once a month.	T. 6	3
3 The efficiency of different techniques used to detect, locate and repair sources of real losses is analyzed and compared for each sector or zone in which uncontrolled water balances are calculated.	T. 2	1
4 Performance levels and reference parameters are determined in order to guide practice and scope of identifying and reducing real losses (evaluation and tracking at least once a year).	T. 2	2
5 Real loss reduction is one of the considerations and objectives of infrastructure renewal and pressure management policies.	T. 2	2
6 References and records of water loss "incidents" are available in geographical databases.	T. 6	2
7 Evaluation of real losses for the geographical area to be rated is based, at the very least, on balance contrasting and minimum flow rates for the entire geographical area or for the sum of smaller areas that make up the geographical area to be rated.	T. 6	1

8	Reliability indicators exist for measurements of minimum night flow rates supplied to sectors or to points where they are recorded and used for loss management.	T. 2	1
9	Procedures exist for monitoring (at least once a day) fluctuations in average and minimum flow rates at sector level to support loss reduction actions.	T. 6	1

OE1.4 Real losses in the water supply, transportation and distribution infrastructure

Real losses refer to the volume of water exiting the supply and distribution infrastructure unintentionally and at unplanned network points without an established use or purpose.

Definition: Daily volume of physical water losses in the "geographical area to be rated" due to poor supply, transportation and distribution infrastructure condition or operation as a proportion of pipe length or number of service connections in the calendar year preceding the rating date. This volume must be taken into account irrespective of whether losses are attributable to repaired ruptures or hidden underground leaks.

Type: Indicator
Service: Drinking Water
Glossary: Geographical area to be rated
Formula: If connections density < 20
        ([EO1-V3]/[EO1-V4])   Unit: m3/km/day
        If connections density >= 20
        ([EO1-V3]/[EO1-V5])   Unit: m3/connection/day
Normalization Function:
           If connections density < 20

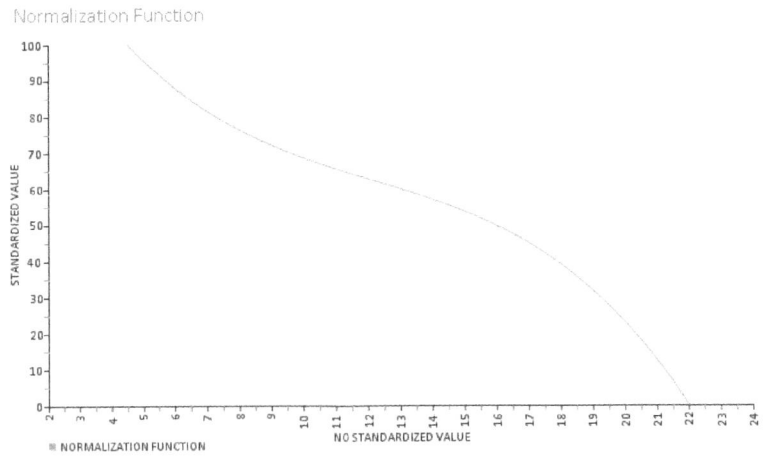

If connections density >= 20

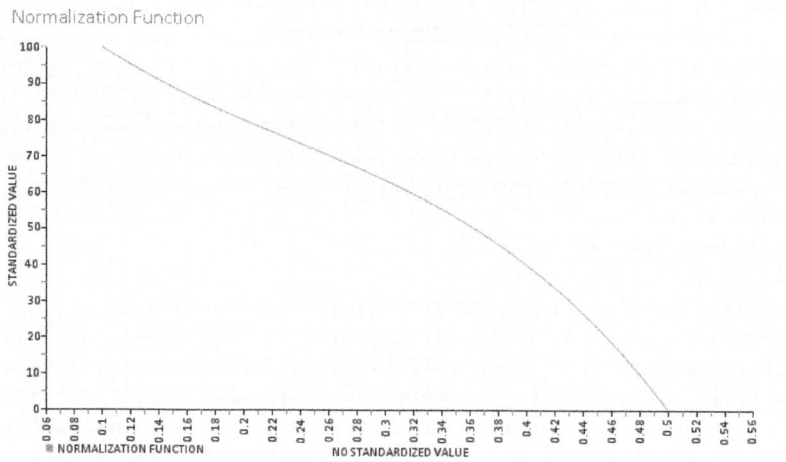

Normalization Function

Variables

[EO1-V3] Volume of water physically lost due to poor supply, transportation and distribution infrastructure condition or operation
Definition: Volume of water physically lost due to poor supply, transportation and distribution infrastructure condition or operation in the calendar year preceding the rating date. It includes losses of both raw and treated water. This volume must be taken into account irrespective of whether losses are attributable to repaired ruptures or hidden underground leaks.
Units: m3/day
Reliability: Table 42

[EO1-V4] Length of supply, transportation and distribution pipes
Definition: Length of supply, transportation and distribution pipes in the "geographical area to be rated" and for operation and maintenance of which the utility is responsible (at the end of the calendar year preceding the rating date). It includes both pipes that transport raw water and those that transport treated water. It excludes the length of service connection pipes.
Units: km
Reliability: Table 43

[EO1-V5] Total number of drinking water service connections at the end of the last full calendar year
Definition: Total number of drinking water service connections at the end of the calendar year preceding the rating date.
Units: connections
Reliability: Table 44

OE1.5 Management of water used in operation

Type: Best Practices
Service: Drinking Water
Normalization: Weighted by practices
Glossary:
Definition: Includes:

Practices	Reliability	Weight
1 A geo-referenced database exists for recording drainage, tank-emptying and filter-washing operations.	T. 3	1
2 A detailed criterion exists for determining volumes lost in each operation based on the length of the operation which loses water and working pressures or on occasional flow rate measurements.	T. 6	1
3 A system exists for recording substitution or installation of new infrastructure that enables evaluation of the water used in infrastructure start-up.	T. 3	1
4 A procedure or plan exists for explicit reduction of water used in operation.	T. 6	2
5 A system exists for contrasting expected and real flow rates at zone or sector level and is used to validate operational water balances.	T. 6	1

OE1.6 Water used in operation

Assesses volumes used in supply and distribution "system" operations, such as water not reclaimed when cleaning treatment plant filters, facility pipe cleaning and repair work, occasional or systematic purges to assure appropriate water quality, and tank emptying for cleaning and maintenance.

It is assessed by comparing volumes used in operation and total "volume introduced into the system".

Definition: Percentage of water used voluntarily and intentionally in operation of the supply, treatment and distribution infrastructure as a proportion of total "water introduced into the system" in the calendar year preceding the rating date.
Type: Indicator
Service: Drinking Water
Glossary: System, Water volume incorporated into the system
Formula: ([EO1-V6]/[EO1-V2])*100   Unit: %
Normalization Function:

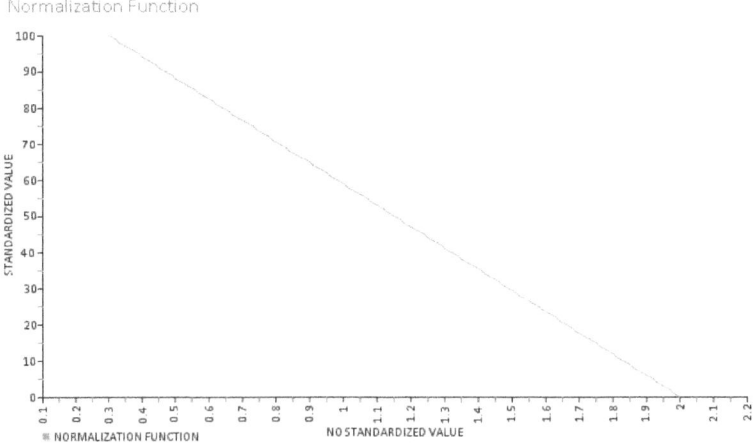

Variables

[EO1-V2] Total "water volume introduced into the system"
Definition: Total "water volume introduced into the system"
Units: m3
Reliability: Table 41

[EO1-V6] Water volume used in infrastructure operation
Definition: Water volume used in infrastructure operation (includes purges and cleaning of pipes, tanks, equipment and facilities in general carried out voluntarily by the utility).
Units: m3
Reliability: Table 45

OE1.7 Management of "reclaimed water"

Type: Best Practices
Service: Drinking Water and/or Sanitation
Normalization: Weighted by practices
Glossary: Reclaimed water
Definition: Includes:

Practices	Reliability	Weight
1 A plan is in force for direct reuse of wastewater.	T. 2	1
2 A differentiated tariff for "reclaimed water" is applied.	T. 2	1
3 Incentivization measures are applied to promote the use of "reclaimed water" in the public and private spheres.	T. 2	1
4 A system exists to monitor "reclaimed water" quality.	T. 2	1
5 A set of rules, regulations and standards specific to "reclaimed water" infrastructure exists.	T. 2	1

OE1.8 "Reused water"

Assesses the degree of direct reuse of wastewater reclaimed in specific plants after treatment to comply with the quality requirements enabling the use for which it is destined. Only wastewater volumes reclaimed and reused in the geographical area served by the rated utility will be assessed. Wastewater volumes reclaimed (and reused) on private property (industrial, institutional or domestic) will not be considered. This indicator is included in this rating area as reclaimed water supply is not considered in the service quality area because it is not yet a widespread activity.

Optimal (efficient) reuse of water within each system depends on multiple factors, which would make assessment of the degree of reuse in each case highly complex. However, it is assumed that a certain level of reuse will always be indicative of efficient management of water resources.

It will be measured by comparing the known volumes of reclaimed water and the total "volumes introduced into the system" for treatment and distribution as fit for consumption.

Definition: Percentage of "reclaimed" water "reused" within the assessed utility's "geographical area to be rated" as a proportion of total "water introduced into the system" for treatment and distribution for consumption in the full calendar year preceding the rating.
Type: Indicator
Service: Drinking Water and/or Sanitation
Glossary: Water volume incorporated into the system, Reclaimed water, Reused water, Geographical area to be rated
Formula: $([EO1-V7]/[EO1-V2])*100$ Unit: %
Normalization Function:

OE Operating Efficiency

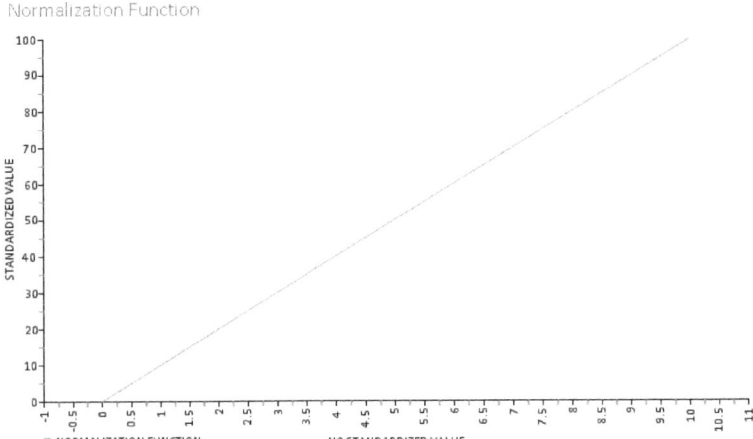

Variables

[EO1-V2] Total "water volume introduced into the system"
Definition: Total "water volume introduced into the system"
Units: m3
Reliability: Table 41

[EO1-V7] Volume of "reclaimed water"
Definition: Volume of "reclaimed water" used within the "geographical area to be rated".
Units: m3
Reliability: Table 46

OE2 Energy usage efficiency

Energy usage in drinking water and wastewater services depends greatly on regulatory requirements governing water supply quality and the quality of wastewater treatment plant effluent. It also depends on the characteristics of the raw water introduced into the "system", the zone's orography, and the type of activity that discharges its wastewater into the collection and treatment network. Nevertheless, the following assessment elements which enable assessment of energy usage efficiency are considered pertinent.

Practices

OE2.1   Energy usage efficiency

Indicators

OE2.2   Energy use in reducing pollutant load

OE2.1 Energy usage efficiency

Type: Best Practices
Service: Drinking Water and/or Sanitation
Normalization: Weighted by practices
Glossary: System
Definition: Includes:

Practices		Reliability	Weight
1	Energy audits that include all energy-consuming facilities in the "system" are carried out at least once every five years.	T. 47	3
2	Measures and recommendations proposed in energy audits are implemented, at least in facilities that account for 90% of total recommendations as measured by energy consumption.	T. 47	3
3	Plans exist for optimizing energy consumption in operation of drinking water supply, treatment, and distribution "systems" and in operation of wastewater collection and treatment systems.	T. 2	2
4	Energy optimization is considered during the infrastructure and equipment design phase.	T. 2	2
5	Energy optimization is considered when planning operation of facilities and of the "system" as a whole.	T. 2	1
6	A plan exists for improving and reducing unit energy consumption and includes annual objectives and monitoring of objective fulfilment.	T. 2	2

OE2.2 Energy use in reducing pollutant load

An indicator (energy use per unit of pollutant load reduced in wastewater treatment plants) is used to complement assessment of energy use in wastewater treatment processes. Energy consumption is not the only variable that determines efficiency in wastewater treatment processes, although it does generally have the most influence on the relationship between pollutant load inflow and the outflow load discharged into the natural environment after treatment. This is why it has been chosen as a representative assessment element with the potential to achieve consistent rating across the systems implemented.

Definition: Energy consumption of all wastewater treatment processes per kilogram of BOD5 pollutant load reduced between inflow and outflow. The average value for the full calendar year preceding the rating will be used.
Type: Indicator
Service: Sanitation
Glossary:
Formula: [EO2-V1]/[EO2-V2]   Unit: kwh/kg BOD5
Normalization Function:

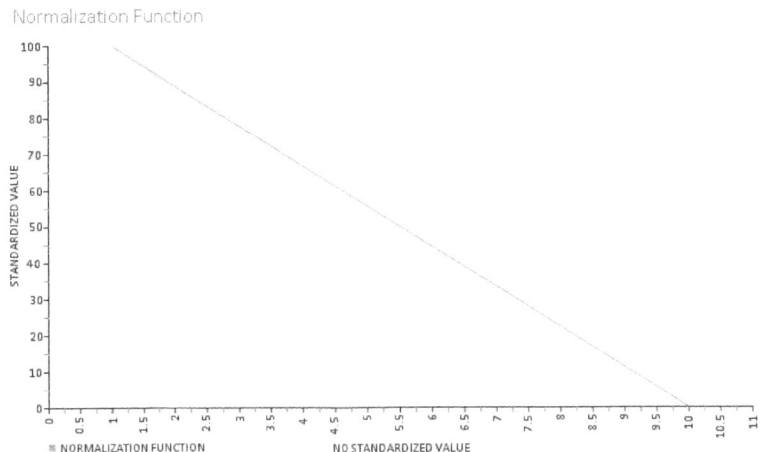

Normalization Function

Variables

[EO2-V1] Total energy consumption in all operative wastewater treatment plants
Definition: Total energy consumption in all wastewater treatment plants operating throughout the reference year.
Units: kWh
Reliability: Table 48

[EO2-V2] Difference in total annual BOD5 kilograms between inflows to wastewater treatment plants operating throughout the entire reference year and BOD5 kilograms in the corresponding outflows.
Definition: Difference in total annual BOD5 kilograms between inflows to wastewater treatment plants operating throughout the entire reference year and BOD5 kilograms in the corresponding outflows.
Units: kg BOD5
Reliability: Table 49

OE3 Infrastructure management efficiency

Infrastructure operating and maintenance efficiency can be measured by the number of malfunctions that occur fortuitously and by the time taken to resolve them. In many cases, these malfunctions interrupt service provision, which is already assessed in the Service Quality area. Nevertheless, it seems appropriate to include in this operating efficiency assessment area parameters which enable specific rating of preventive and corrective infrastructure maintenance tasks in general and that quantify the inputs allocated and the results achieved.

This section does not include considerations linked to planning effort in infrastructure management, which has been taken into account in the Planning Efficiency area. Here, it is limited to inspection tasks and their ensuing corrective actions, as well as to contingency and malfunction resolution as part of corrective and preventive maintenance.

Practices

OE3.1 Efficiency in management of water intake, treatment and distribution infrastructure

OE3.6 Efficiency in management of wastewater collection and treatment infrastructure

Indicators

OE3.2 Number of ruptures in transportation and distribution pipes

OE3.3 Number of ruptures in service connections (connections up to private supply systems)

OE3.4 Expenditure on "corrective maintenance" of fixed physical assets linked to the water intake, treatment and distribution "system"

OE3.5 Expenditure on "preventive maintenance" of fixed physical assets linked to the water intake, treatment and distribution "system"

OE3.7 Fortuitous "incidents" affecting the wastewater collection network during dry weather

OE3.8 Expenditure on "corrective maintenance" of fixed physical assets linked to the wastewater collection and treatment "system".

OE3.9 Expenditure on "preventive maintenance" of fixed physical assets linked to the wastewater collection and treatment "system"

OE3.1 Efficiency in management of water intake, treatment and distribution infrastructure

Type: Best Practices
Service: Drinking Water
Normalization: Weighted by practices
Glossary: System, Incident, Preventive maintenance, Corrective maintenance
Definition: Includes:

	Practices	Reliability	Weight
1	Geo-referenced databases (GIS) for all infrastructure are available.	T. 3	1
2	A unit is specifically responsible for maintaining and updating infrastructure information in the GIS.	T. 6	1
3	Procedures are used to ensure updating of information in the GIS regarding infrastructure characteristics and include a commitment to update within a stipulated timeframe.	T. 6	1
4	A remote control system exists that relays the operational status of at least 20% of maneuverable devices and equipment positioned in strategic parts of the "system".	T. 3	1
5	An early warning system exists to identify "incidents" (remote control, sectoring, online indicators).	T. 3	2
6	An integrated system exists to manage inspection and "preventive maintenance".	T. 3	2
7	A system to manage pressure in the distribution network is available and implemented.	T. 3	2
8	An integrated system exists to manage anomaly reporting and resolution linked to the operation and warnings and complaints areas.	T. 3	3
9	A "preventive maintenance" plan exists.	T. 6	1
10	"Preventive" and "corrective" maintenance costs are monitored and controlled.	T. 6	1

A line of research or analysis is conducted on the
11 performance and service life of the equipment and T. 6 1
infrastructure.

OE3.2 Number of ruptures in transportation and distribution pipes

Definition: Annual number of reported ruptures in transportation or distribution pipes per kilometer of such pipes in the "system". The average for the last 3 years will be taken.
Type: Indicator
Service: Drinking Water
Glossary: System, Geographical area to be rated
Formula: [EO3-V1]/[EO1-V4]   Unit: N°/km
Normalization Function:

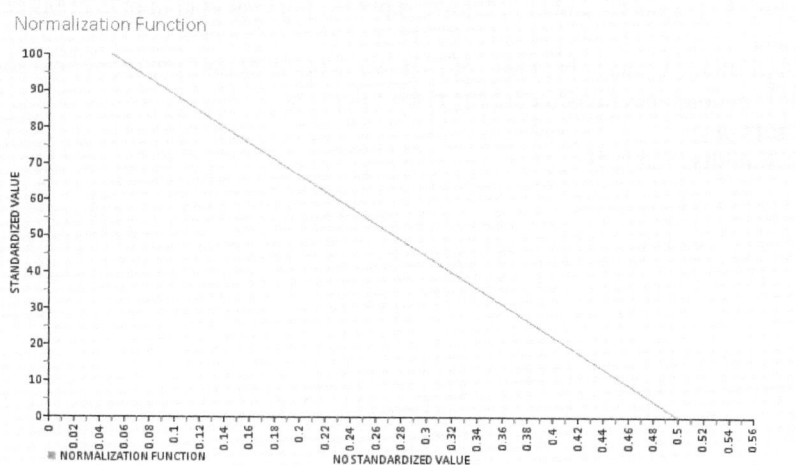

Variables

[EO1-V4] Length of supply, transportation and distribution pipes
Definition: Length of supply, transportation and distribution pipes in the "geographical area to be rated" and for operation and maintenance of which the utility is responsible (at the end of the calendar year preceding the rating date). It includes both pipes that transport raw water and those that transport treated water. It excludes the length of service connection pipes.
Units: km
Reliability: Table 43

[EO3-V1] Annual number of known ruptures in transportation or distribution pipes
Definition: Annual number of known ruptures in transportation or distribution pipes (average for the last 3 full years).
Units: no.
Reliability: Table 50

OE3.3 Number of ruptures in service connections (connections up to private supply systems)

Definition: Number of reported ruptures per 100 connections (average for the last 3 full years).
Type: Indicator
Service: Drinking Water
Glossary:
Formula: ([EO3-V2]/[EO1-V5])*100   Unit: N°/100 connections
Normalization Function:

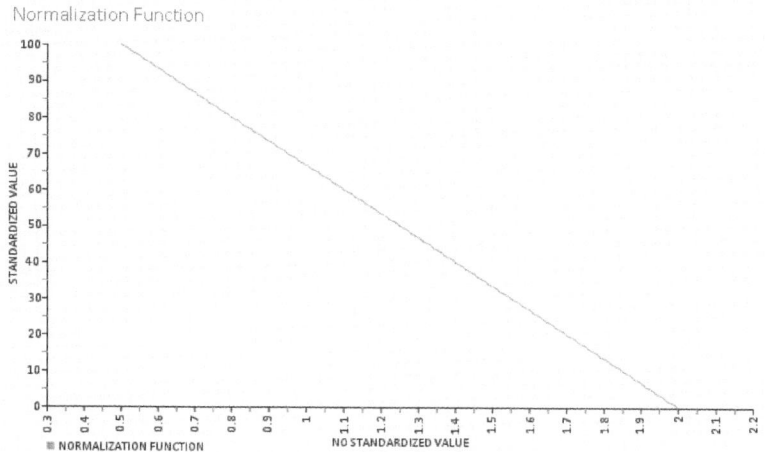

Variables

[EO1-V5] Total number of drinking water service connections at the end of the last full calendar year
Definition: Total number of drinking water service connections at the end of the calendar year preceding the rating date.
Units: connections
Reliability: Table 44

[EO3-V2] Annual number of known ruptures in connections
Definition: Annual number of known ruptures in connections (average for the last 3 full years)
Units: no.
Reliability: Table 51

OE3.4 Expenditure on "corrective maintenance" of fixed physical assets linked to the water intake, treatment and distribution "system"

Considers expenditure on all "corrective maintenance" performed on the water intake, treatment and distribution "system", including "incident" resolution, as a proportion of the value of the corresponding assets. It includes repair of ruptures, as well as all other incidents that affect service. Infrastructure replacements not programmed in the renewal plans will be counted as corrective maintenance. Compensation costs for damages due to anomalies will also be included. If a specific policy exists, the annual cost of the policy will be added. Expenses for the last full calendar year and asset values at the beginning of the financial period will be considered.

Definition: Percentage representing annual expenditure on "corrective maintenance" of the fixed physical assets linked to the water intake, treatment and distribution "system" as a proportion of their gross value (except land) at the beginning of the last full calendar year. The average for the last 3 full calendar years is considered.
Type: Indicator
Service: Drinking Water
Glossary: System, Incident, Corrective maintenance, Geographical area to be rated
Formula: ([EO3-V3]/[EP3-V2.1])*100   Unit: %
Normalization Function:

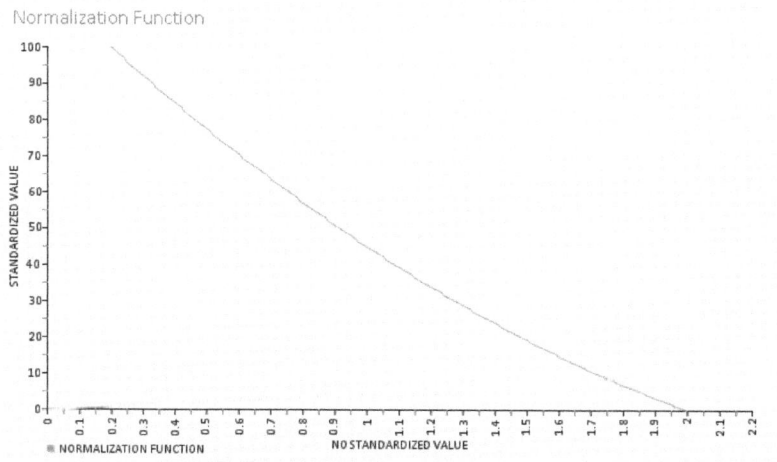

Variables

[EO3-V3] Total annual expenditure on all "corrective maintenance" performed on water intake, treatment and distribution "systems"
Definition: Total annual expenditure on all "corrective maintenance" performed on water intake, treatment and distribution "systems", including incidents resolved, replacement not programmed in renewal plans, compensation for damages to third parties, and specific insurance policies.
Units: financial statement currency
Reliability: Table 35

[EP3-V2.1] Gross value of the fixed physical assets linked to the water intake, treatment and distribution "system"
Definition: Gross value of the facilities, equipment and infrastructure linked to the water intake, treatment and distribution "systems" in the "geographical area to be rated" (except land), including infrastructure not owned by the utility in the case that the utility is responsible for its replacement and maintenance costs. The gross value must match the gross value entered in the accounts at the beginning of the year, including value adjustments if applicable.
Units: financial statement currency
Reliability: Table 36

OE3.5 Expenditure on "preventive maintenance" of fixed physical assets linked to the water intake, treatment and distribution "system"

Total expenditure on "preventive maintenance" performed on the water intake, treatment and distribution "system", including inspection, handling and resolution of anomalies detected during inspection, and replacement as a proportion of the gross value of the corresponding fixed physical assets (except land). In neither of these two cases are the costs of programmed renewal of infrastructure and facilities considered.

Definition: Percentage representing annual expenditure on "preventive maintenance" of the fixed physical assets linked to the water intake, treatment and distribution "system" as a proportion of their gross value (except land) at the beginning of the last full calendar year. The average for the last three full calendar years is considered.
Type: Indicator
Service: Drinking Water
Glossary: System, Preventive maintenance, Geographical area to be rated
Formula: ([EO3-V4]/[EP3-V2.1])*100   Unit: %
Normalization Function:

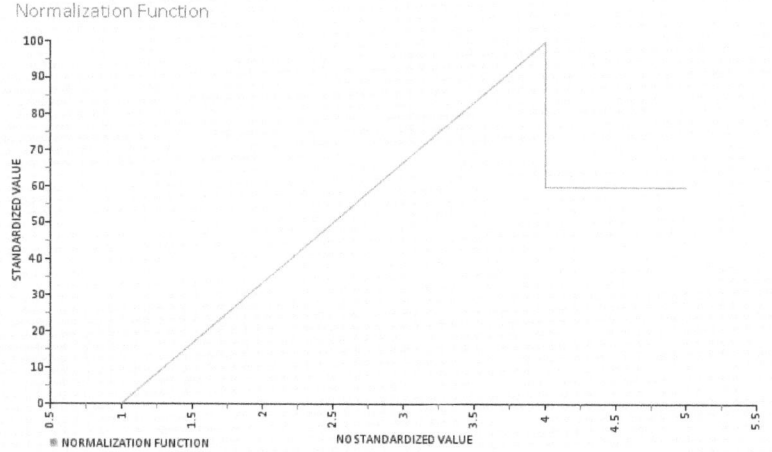

Variables

[EO3-V4] Annual expenditure on "preventive maintenance" of fixed physical assets linked to the water intake, treatment and distribution "system"
Definition: Annual expenditure on "preventive maintenance" of fixed physical assets linked to the water intake, treatment and distribution "system" (last 3 full calendar years).
Units: financial statement currency
Reliability: Table 35

[EP3-V2.1] Gross value of the fixed physical assets linked to the water intake, treatment and distribution "system"
Definition: Gross value of the facilities, equipment and infrastructure linked to the water intake, treatment and distribution "systems" in the "geographical area to be rated" (except land), including infrastructure not owned by the utility in the case that the utility is responsible for its replacement and maintenance costs. The gross value must match the gross value entered in the accounts at the beginning of the year, including value adjustments if applicable.
Units: financial statement currency
Reliability: Table 36

OE3.6 Efficiency in management of wastewater collection and treatment infrastructure

Type: Best Practices
Service: Sanitation
Normalization: Weighted by practices
Glossary: System, Incident, Preventive maintenance, Corrective maintenance
Definition: Includes:

	Practices	Reliability	Weight
1	Geo-referenced databases (GIS) for all infrastructure are available.	T. 3	1
2	A unit is specifically responsible for maintaining and updating infrastructure information in the GIS.	T. 6	1
3	Procedures are used to ensure updating of information regarding infrastructure characteristics and include a commitment to update within a stipulated timeframe.	T. 6	1
4	A remote control or equivalent online system exists to relay the operational status of maneuverable devices and equipment in the wastewater or drainage network. (In "systems" with networks that do not include maneuverable devices or equipment, this practice is considered complied with at maximum reliability level).	T. 3	1
5	An early warning system exists to identify "incidents" (remote control, sectoring, online indicators).	T. 3	2
6	An integrated system exists to manage inspection and "preventive maintenance".	T. 3	2
7	An integrated system exists to manage anomaly reporting and resolution linked to the operation and warnings and complaints areas.	T. 3	3
8	A "preventive maintenance" plan exists.	T. 6	1
9	"Preventive" and "corrective" maintenance costs are monitored and controlled.	T. 6	1

A line of research or analysis is conducted on the
10 performance and service life of the equipment and   T. 6   1
infrastructure.

OE3.7 Fortuitous "incidents" affecting the wastewater collection network during dry weather

Definition: Number of "incidents" per thousand kilometers of wastewater collection network in the last full calendar year. Network length considers all sewers and sewer mains up to discharge into wastewater treatment plants, small scale solutions or the receiving environment. Marine outfalls will not be considered (neither incidents nor length).
Type: Indicator
Service: Sanitation
Glossary: Incident, Geographical area to be rated
Formula: ([EO3-V5]/[EO3-V6])*1000   Unit: Nº/1000km
Normalization Function:

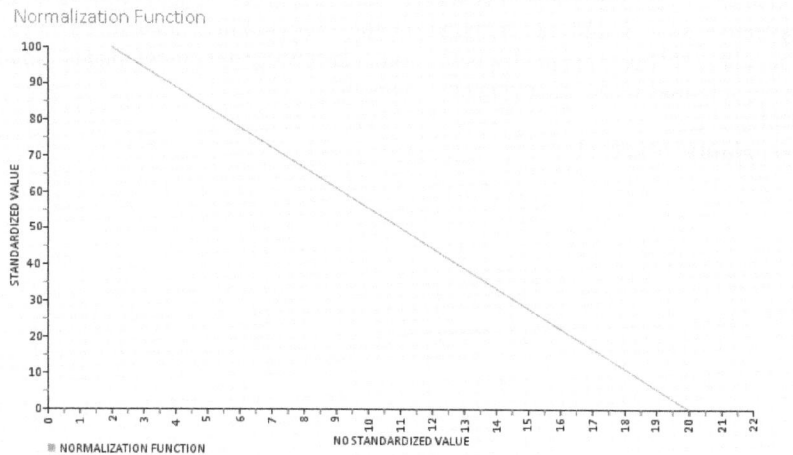

Variables

[EO3-V5] Number of "incidents" affecting the wastewater collection network during dry weather
Definition: Average daily number of "incidents" affecting the wastewater collection network during dry weather in the full calendar year preceding the rating date. All anomalies requiring intervention to remedy them and that threaten proper operation of the wastewater collection network are considered incidents.
Units: no.
Reliability: Table 52

[EO3-V6] Length of wastewater collection network
Definition: Length of wastewater collection network at the end of the calendar year preceding the rating date. Network length considers all operative sewers and sewer mains in the "geographical area to be rated", including sections up to discharge into wastewater treatment plants, small scale solutions or the receiving environment.
Units: km
Reliability: Table 43

OE3.8 Expenditure on "corrective maintenance" of fixed physical assets linked to the wastewater collection and treatment "system".

Considers expenditure on all "corrective maintenance" performed on the wastewater collection and treatment "system", including "incident" resolution, as a proportion of the value of the corresponding assets. It includes repair of ruptures, as well as all other incidents that affect service. Infrastructure replacements not programmed in the renewal plans will be counted as corrective maintenance. Compensation costs for damages due to anomalies will also be included. If a specific policy exists, the annual cost of the policy will be added. Expenses for the last full calendar year and asset values at the beginning of the financial period will be considered.

Definition: Percentage representing annual expenditure on "corrective maintenance" of fixed physical assets linked to the wastewater collection and treatment "system" as a proportion of their gross value (except land) at the beginning of the last full calendar year. The average for the last 3 full calendar years is considered.
Type: Indicator
Service: Sanitation
Glossary: System, Incident, Corrective maintenance, Geographical area to be rated
Formula: ([EO3-V7]/[EP3-V2.2])*100   Unit: %
Normalization Function:

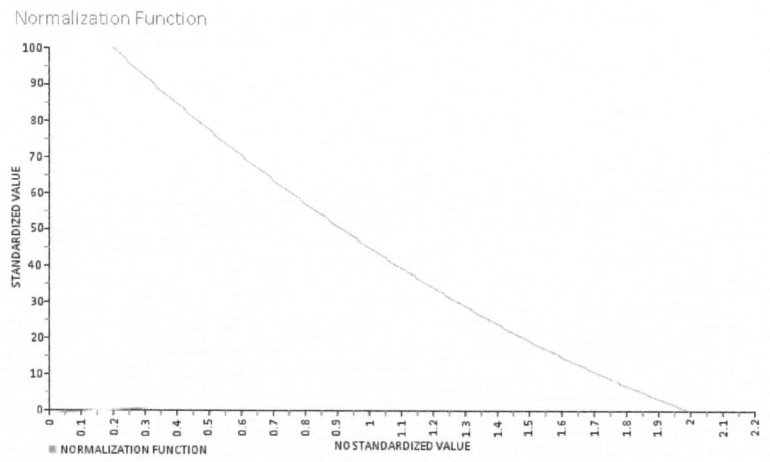

Normalization Function

STANDARDIZED VALUE

■ NORMALIZATION FUNCTION   NO STANDARDIZED VALUE

Variables

[EO3-V7] Total annual expenditure on "corrective maintenance" of the wastewater collection and treatment "system"
Definition: Total annual expenditure on "corrective maintenance" of the wastewater collection and treatment "system". Includes incident resolution, replacement not programmed in renewal plans, compensation for damages to third parties, and specific insurance policies.
Units: financial statement currency
Reliability: Table 35

[EP3-V2.2] Gross value of fixed physical assets linked to the wastewater collection and treatment "system"
Definition: Gross value of the facilities, equipment and infrastructure linked to the wastewater collection and treatment "system" in the "geographical area to be rated" (except land), including infrastructure not owned by the utility, in the case that the utility is responsible for its replacement and maintenance costs. The gross value must match the gross value entered in the accounts at the beginning of the year, including value adjustments if applicable.
Units: financial statement currency
Reliability: Table 36

OE3.9 Expenditure on "preventive maintenance" of fixed physical assets linked to the wastewater collection and treatment "system"

Total expenditure on "preventive maintenance" performed on the wastewater collection and treatment "system", including inspection, handling and resolution of anomalies detected during inspection, and replacement as a proportion of the gross value of the corresponding fixed physical assets (except land). Costs of programmed renewal of infrastructure and facilities are not considered.

Definition: Percentage representing annual expenditure on "preventive maintenance" of fixed physical assets linked to the wastewater collection and treatment "system" as a proportion of their gross value (except land) at the beginning of the last full calendar year. The average for the last 3 full calendar years is considered.
Type: Indicator
Service: Sanitation
Glossary: System, Preventive maintenance, Geographical area to be rated
Formula: ([EO3-V8]/[EP3-V2.2])*100   Unit: %
Normalization Function:

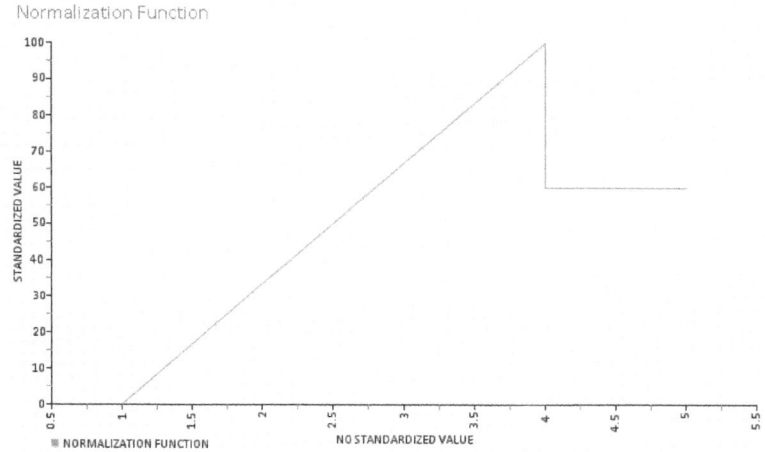

Variables

[EO3-V8] Annual expenditure on "preventive maintenance" of fixed physical assets linked to the wastewater collection and treatment "system"
Definition: Annual expenditure on "preventive maintenance" of fixed physical assets linked to the wastewater collection and treatment "system"
Units: financial statement currency
Reliability: Table 35

[EP3-V2.2] Gross value of fixed physical assets linked to the wastewater collection and treatment "system"
Definition: Gross value of the facilities, equipment and infrastructure linked to the wastewater collection and treatment "system" in the "geographical area to be rated" (except land), including infrastructure not owned by the utility, in the case that the utility is responsible for its replacement and maintenance costs. The gross value must match the gross value entered in the accounts at the beginning of the year, including value adjustments if applicable.
Units: financial statement currency
Reliability: Table 36

OE4 Operational and maintenance cost efficiency

Operating costs are the element that best characterizes operating efficiency if expressed in relation to the quantity of water supplied or treated or to the number of users or size of population served. However, as cases differ widely it is not possible to make a balanced assessment unless complex homogenization is performed to take into account the multiple context variables that condition the values obtained. Therefore, use of an assessment based on practices is proposed that makes it possible to segregate incurred costs, analyze efficiency levels and verify the existence of optimization plans for those costs.

Practices

OE4.1   Operational and maintenance cost efficiency

OE4.1 Operational and maintenance cost efficiency

Type: Best Practices
Service: Drinking Water and/or Sanitation
Normalization: Weighted by practices
Glossary: System
Definition: Includes:

	Practices	Reliability	Weight
1	Individual annual accounting analyses are carried out for operating and maintenance costs as a whole.	T. 2	1
2	Individual annual accounting analyses are carried out for "system" operating and maintenance costs.	T. 2	1
3	Individual monthly accounting analyses are carried out for main system facilities. These comprise, at the very least, all water treatment plants, wastewater treatment plants, main pumping stations, water distribution and wastewater collection networks and "systems" supporting operation.	T. 6	2
4	Individual and segregated monthly accounting analyses are carried out for the main components of these operating costs: staff, reagents, energy consumption and third-party services.	T. 6	2
5	Operating cost optimization is included in the infrastructure and equipment design phase.	T. 6	2
6	Operating cost optimization is considered when planning facility operation and operation of the "system" as a whole.	T. 6	2
7	A plan exists to reduce operating unit costs and includes annual objectives and monitoring of their fulfilment.	T. 6	3

ME Business Management Efficiency

Business management efficiency is essential to organizations' long-term competitiveness and sustainability. Assessment of this area firstly considers the utility's strategic plan from the point of view of its content, formulation process and implementation plan. The second sub-area assesses whether a "management control system" is in place to monitor and control the utility's performance. The third one considers assessment of the organizational structure's main characteristics, including whether an updated organization chart exists, whether basic functions related to supply of drinking water and wastewater services are considered and whether functions, authority and profiles for each position are clearly defined. The fourth sub-area assesses the management aspects considered to be most relevant to recruiting, appraising, developing and retaining the human capital required to fulfil the utility's objectives. The fifth one considers procurement management from the point of view of its efficacy, compliance with standards currently in force, participation and transparency. Finally, the sixth sub-area is cross-cutting and assesses staff efficiency and efficiency in use of support resources.

The sub-areas of evaluation are:

ME1 Strategic planning
ME2 Management control
ME3 Organizational structure
ME4 Human resource management
ME5 Procurement management

ME6 Staff and support resource efficiency

ME1 Strategic planning

___

One of the elements needed to achieve efficient management in an organization is the existence and implementation of a strategic plan, understanding by this an integrated set of actions designed to achieve the institution's medium/long-term objectives and goals, together with allocation of the resources required. The strategic plan establishes objectives and goals, specifies the policies and lines of action with which to achieve those objectives and defines concrete and explicit timeframes, which must be fulfilled in order for the plan's implementation to be successful.

This sub-area assesses the content of the utility's strategic plan, its formulation process and its implementation based on two groups of good practices.

Practices

ME1.1   Strategic plan contents
ME1.2   Strategic plan formulation and implementation

ME1.1 Strategic plan contents

Type: Best Practices
Service: Drinking Water and/or Sanitation
Normalization: Weighted by practices
Glossary: Strategic map
Definition: A strategic plan exists containing the following:

Practices	Reliability	Weight
1 A utility mission or rationale and vision of the future exist.	T. 53	1
2 A summary exists of the strengths, weaknesses, opportunities and threats identified in relation to key aspects of performance, such as access to service, service quality, management efficiency, environmental sustainability and financial sustainability. At the very least, an analysis of critical success factors exists.	T. 53	1
3 The utility's strategic position is analyzed with respect to the opportunities and threats identified, or with respect to the critical success factors, depending on the methodology used for the previous practice.	T. 53	2
4 The utility's situation is analyzed with respect to objectives and goals considered in the sector or local development plan at the respective governmental level, in the case that this plan is documented and applies to the utility.	T. 53	1

Analysis/assessment exists of strategic options to improve the utility's strategic position and/or to overcome detected weaknesses, considering, among other things, at least 2 of the following options:
(i) infrastructure construction through BOT or similar contracts;
(ii) organizational restructuring;
(iii) outsourcing of some services;
(iv) outsourcing of management of a geographical area, of any of the basic functions of service delivery (such as sales or maintenance functions), or of the institution as a whole, by means of management contracts. (This practice is considered as complied with at maximum reliability level if (i) the strategic options analyzed include just 1 of the

5	aforementioned options, but at least 2 of them are being implemented, or (ii) none of the aforementioned options are analyzed, but at least 3 of them are being implemented).	T. 53	4
6	Strategic objectives (determining intended impact and direction) are defined for a horizon of at least 5 years for the key performance areas prioritized as a result of analysis of strategic options.	T. 53	3
7	Specific objectives associated with each of the strategic objectives, with their respective goals and deadlines, are identified.	T. 53	5
8	Strategic guidelines are defined regarding customers, services offered, stakeholders, processes, staff and technology.	T. 53	2
9	Each of the activities considered in the plan is budgeted.	T. 53	3
10	A "map" exists of the strategy selected.	T. 53	1

ME1.2 Strategic plan formulation and implementation

Type: Best Practices
Service: Drinking Water and/or Sanitation
Normalization: Weighted by practices
Glossary: Board of directors
Definition: Includes:

Practices	Reliability	Weight
1 The plan was generated by utility staff (at the very least involving management staff up to the third hierarchical level, if this level exists in the organization) through a formal participatory process.	T. 54	2
2 The strategic plan is reviewed annually.	T. 55	1
3 The process of reviewing the plan includes assessment of the degree of success achieved in its implementation and analysis of the environment (natural, social, political, economic, regulatory) in which services are provided in order to detect possible changes to conditions under which the plan was formulated, analyze their possible impact on strategy and determine the need for adjustment.	T. 55	1
4 The strategic plan has been formally approved by the "board of directors".	T. 27	1
5 Operating plans derived from the strategy are specified for functional areas, departments and all units.	T. 56	3
6 Special projects identified to ensure fulfilment of the strategic plan's objectives are included, indicating their objective, duration, costs and person or persons in charge of each project.	T. 56	1
7 A plan for monitoring the strategic plan's implementation is included, specifying responsibilities for delivery.	T. 57	2

A program exists for communicating the plan to all levels and staff in the organization so that it is clear to them what the organization intends to achieve, the guidelines to be followed to achieve it and their specific contribution to

| 8 | fulfilment of the goals and to performance of the activities for which the unit in which they work is responsible. | T. 57 | 3 |

| 9 | Measures, mechanisms and criteria exist for assessing the strategy in order to provide means of assessing the plan's degree of success. | T. 56 | 1 |

ME2 Management control

The existence of a "management control system" is a key factor in monitoring utility performance so that the actions required to ensure accomplishment of the objectives defined are taken in a timely manner.

This criterion's objective is to assess if the utility has a system in place that enables systematic measurement, assessment and monitoring of institutional performance in pursuit of the objectives and goals established either in the strategic plan, in the case that such an instrument exists, or in any other mechanism used to formulate goals.

The following assessment elements are considered measures of good practice: system completeness, assessment systematicity and the degree of accomplishment of the goals established.

Practices

ME2.1 "Management control system"

ME2.1 "Management control system"

Type: Best Practices
Service: Drinking Water and/or Sanitation
Normalization: Weighted by practices
Glossary: Board of directors, Management control system
Definition: Includes:

Practices		Reliability	Weight
1	An annual budget is prepared and used to allocate and control use of resources by the utility. This budget is monitored every month.	T. 58	3
2	Reports on the utility's management are prepared at least once a quarter and include indicators that measure key aspects such as access to service, service quality, management efficiency, environmental sustainability and financial sustainability.	T. 58	1
3	A management "control system" exists that establishes goals for the indicators that measure key aspects such as access to service, service quality, management efficiency, environmental sustainability and financial sustainability, and which allows regular monitoring and evaluation of the degree of goal fulfilment.	T. 3	2
4	A management "control system" exists that allows regular monitoring and assessment of fulfilment of objectives and goals established in the utility's strategic plan. The system is deployed across all organizational levels and units to guarantee fulfilment of objectives and goals.	T. 3	3
5	Indicators considered in the management "control system" are measurable, pertinent (relevant, specific), verifiable and reasonable in number (not more than 20).	T. 59	1
6	80% or more of total annual goals monitored via the management "control system" are fulfilled (goals are considered fulfilled if they achieve between 95% and 120% of the value formally established before the start of the respective year).	T. 60	2

Management progress and results are monitored monthly by
7 senior management and at least quarterly by the utility's   T. 61   3
"board of directors".

ME3 Organizational structure

Organizational structure (the form in which activities necessary to fulfil the utility's objectives are divided and arranged) is one of the elements that influence business management efficiency.

Although a single or ideal organizational structure does not exist, certain elements of good practice are assessed here, such as existence of an updated organizational chart; organizational user focus; existence of basic functions related to provision of drinking water and wastewater services; and clearly defined functions, authority and job descriptions for each position.

Practices

ME3.1   Organizational structure

ME3.1 Organizational structure

Type: Best Practices
Service: Drinking Water and/or Sanitation
Normalization: Weighted by practices
Glossary: Board of directors
Definition: Includes:

Practices	Reliability	Weight
1	An organizational chart exists that reflects all functional and territorial levels, areas and units which make up the current organizational structure (implemented).	T. 62	1
2	The current organizational structure is documented in a handbook that contains the functions of the different units, general qualifications for positions and the functions, responsibilities, authority and job descriptions for each position.	T. 63	2
3	The current organizational structure meets the following conditions: (i) it is consistent with the role and objectives defined in the applicable legal provisions and/or in the institution's articles of incorporation and bylaws (i.e. the functions of the organizational structure reflect the functions and objectives established in the legal documents that created the organization), and (ii) it includes the changes defined in the strategic plan, in the case that modifications to the organizational structure have been defined in the plan.	T. 56	2
4	The current organizational structure is user-oriented (user or customer support or service units are located at the second or third level of the organizational structure and have the authority to provide service information to users; to receive, answer and resolve standard requests and problems related to these services; and to manage and monitor related processes, such as service availability requests, service connection and disconnection, and billing, among others) and/or based in business units (it is divided or structured into units that can be measured by their economic income - revenue minus expenses).	T. 56	3

All basic functions inherent to drinking water and wastewater services are considered: user services, operations (production, distribution, collection, and treatment, according
5 to the services to be rated), investment project planning and   T. 56   2
management, as well as support functions such as finance, human resources, management control, internal audit, information technology and procurement.

A legal counsel function is considered, responsible for providing counsel on all matters required for timely compliance with legal, regulatory and statutory regulations
6 that affect the institution. It regularizes water rights; conducts   T. 56   1
and follows up all legal actions that affect the institution and all claims that it pursues; and draws up or reviews contracts and agreements required by the institution.

The current organizational structure has been approved by
7 the "board of directors", at least to the third level of the   T. 27   1
organization, considering as first level general management.

ME4 Human resource management

Human resource management is a key factor in any organization, as people are the fundamental and decisive element in business performance and management quality.

This criterion seeks to assess management aspects considered most relevant to recruiting, appraising, developing and retaining the human capital required to achieve the institution's objectives.

Practices

ME4.1 Human resource management

Indicators

ME4.2 Staff recruited competitively
ME4.3 Staff receiving training
ME4.4 Staff who comply with "key position" "job descriptions"

ME4.1 Human resource management

Type: Best Practices
Service: Drinking Water and/or Sanitation
Normalization: Weighted by practices
Glossary: Job description
Definition: Includes:

Practices	Reliability	Weight
1	A human resources information system exists that, besides personal data and remunerations, contains information on staff skill profiles, qualifications, age, experience, years of service, training, and ratings (results of staff performance appraisals).	T. 3	3
2	A staff performance appraisal system exists and is applied.	T. 2	2
3	A training program has been prepared considering the institution's strategic objectives and staff skill gaps. The training plan contains the types of courses, contents, duration and staff positions or areas of specialization for which training is intended.	T. 2	2
4	Results of implemented training programs are assessed.	T. 4	1
5	A defined remuneration policy that considers incentives for performance and results exists and is applied.	T. 6	3
6	A remuneration scale based on the relative importance of positions and which takes market remuneration as a reference is applied. This scale is based on a specific study carried out for this purpose in the 3 calendar years preceding the rating date, or is based on a remuneration survey, not more than 3 years old at the rating date, of utilities providing regulated public services.	T. 6	2

The annual staff turnover rate (measured as the difference between staff joining and leaving the institution during the respective year divided by 2 as a proportion of total monthly average staff numbers in the same year) lies between 2% and 6%, while annual turnover of senior staff that hold first- and second-level positions does not exceed 20%. In both cases,

7	the average turnover rate for the last 3 calendar years preceding the rating date is considered.	T. 64	1
8	Staff who hold first- and second-level management positions have completed 4 or more years of post-secondary education, have a minimum 5 years' work experience since finishing that education and comply with the specific qualification and experience criteria defined in the respective job descriptions.	T. 65	3
9	Surveys are conducted to gauge staff satisfaction and/or assess workplace atmosphere and staff perceptions of change.	T. 4	1
10	Staffing levels and composition are reviewed at least every 3 years according to the "job descriptions" required to fulfill the institution's objectives and to productivity analysis, all of which is recorded.	T. 66	2
11	The work day is designed so that activities are carried out during regular working hours. Shifts are considered for activities that require attention either at night or 24 hours a day. Use of overtime is limited to extraordinary events or activities that require the use of time outside regular working hours, subject to prior authorization by the unit's head.	T. 67	1
12	Measures are implemented to prevent occupational hazards and professional diseases and to ensure occupational health.	T. 67	2

ME4.2 Staff recruited competitively

Definition: Percentage of staff recruited competitively as a proportion of total staff recruited during the 3 full calendar years preceding the rating date. If there was no recruitment in this period, data from the last year in which staff were recruited should be considered.
Type: Indicator
Service: Drinking Water and/or Sanitation
Glossary:
Formula: ([EG4-V1]/[EG4-V2])*100   Unit: %
Normalization Function:

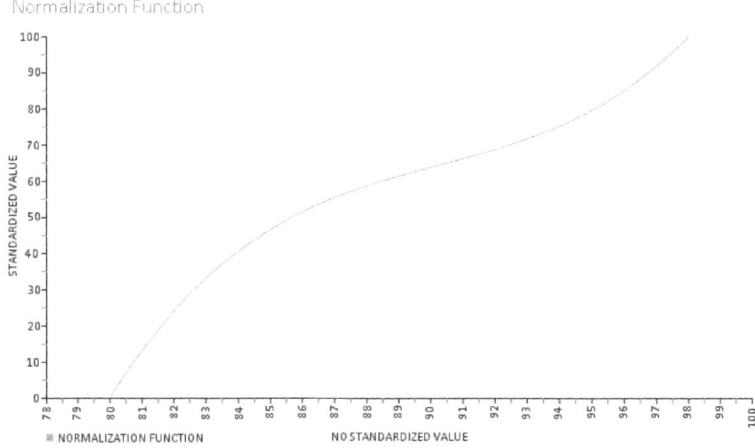

Variables

[EG4-V1] Number of staff recruited competitively
Definition: Number of staff recruited competitively in the 3 full calendar years preceding the rating date. If there was no recruitment in this period, data from the last year in which staff were recruited should be considered.
Units: no.
Reliability: Table 68

[EG4-V2] Total number of staff recruited
Definition: Total number of staff recruited in the 3 full calendar years preceding the rating date. If there was no recruitment in this period, data from the last year in which staff were recruited should be considered.
Units: no.
Reliability: Table 69

ME4.3 Staff receiving training

Definition: Percentage of staff receiving training as a proportion of the total number of utility staff.
Type: Indicator
Service: Drinking Water and/or Sanitation
Glossary: Own staff, Training courses
Formula: ([EG4-V3]/[EG4-V4])*100   Unit: %
Normalization Function:

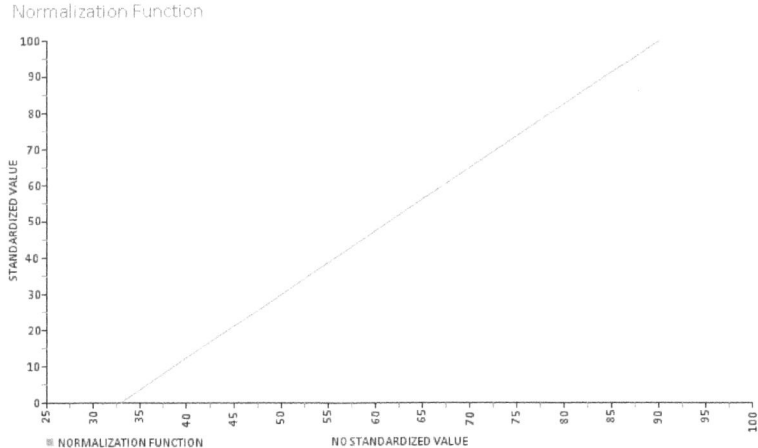

Variables

[**EG4-V3**] Number of "own staff" participating in "training courses"
Definition: Number of "own staff" participating in "training courses" in the full calendar year preceding the rating date.
Units: no.
Reliability: Table 70

[**EG4-V4**] Total staff
Definition: Average total of "own staff" for the full calendar year preceding the rating date.
Units: no.
Reliability: Table 71

ME4.4 Staff who comply with "key position" "job descriptions"

Definition: Percentage of staff holding "key positions" who comply with key position "job descriptions" as a proportion of total staff in key positions at the rating date.
Type: Indicator
Service: Drinking Water and/or Sanitation
Glossary: Key positions, Job description
Formula: ([EG4-V5]/[EG4-V6])*100   Unit: %
Normalization Function:

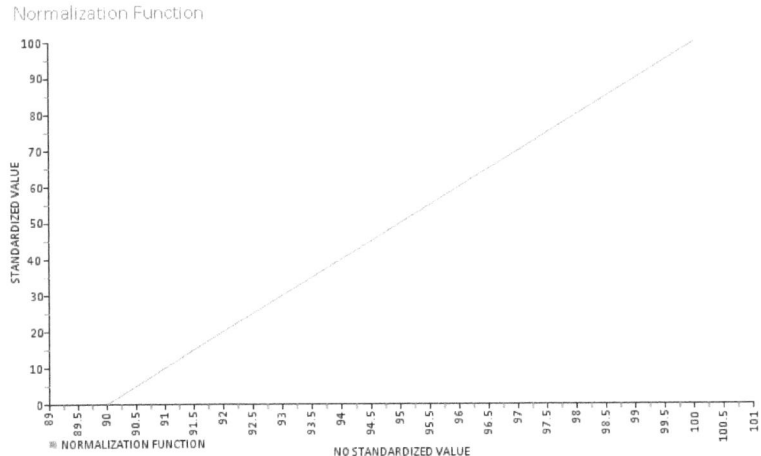

Variables

[EG4-V5] Number of persons holding "key positions" who comply with the respective "job descriptions"
Definition: Number of persons holding "key positions" who comply with the respective "job descriptions"
Units: no.
Reliability: Table 72

[EG4-V6] Total persons who hold "key positions"
Definition: Total persons who hold "key positions"
Units: no.
Reliability: Table 73

ME5 Procurement management

Procurement management has a major impact on both service provision and construction and maintenance of the infrastructure required to deliver water and wastewater services.

This criterion assesses management of procurement of the goods and services required to achieve the institution's objectives, including its effectiveness, compliance with the requirements of regulations in force, participation and transparency. Measures of good practice and three indicators are used as assessment elements.

Practices

ME5.1   Procurement

Indicators

ME5.2   Purchases made by "public tender"
ME5.3   "Successful tenders"
ME5.4   Tenders held within the regulated "minimum timeframe"

ME5.1 Procurement

Type: Best Practices
Service: Drinking Water and/or Sanitation
Normalization: Weighted by practices
Glossary:
Definition: Includes:

Practices	Reliability	Weight
1	A formally defined strategy and policy exists for procurement of goods and services, including studies and works.	T. 6	2
2	A definition exists regarding levels of materials and spare parts' inventory in order to avoid lack of stock and losses due to obsolescence or deterioration, considering associated costs.	T. 6	2
3	New goods, services and alternative supply sources are studied, based on existing options in the market, at least once a year for goods and services which altogether represent more than 70% of the total amount of purchases used in operations in the year preceding the rating date. Options are considered to exist in the market if there is more than one supplier of the goods and services used in operations, considering both national and international markets. In the case that there is only one supplier for a certain good or service, this will be documented.	T. 4	1
4	Purchases of goods and services are made, in accordance with the standards that govern the utility, via an open and transparent online computerized system, through which interested suppliers have access to tender conditions, terms of reference and other conditions which regulate each purchase process.	T. 74	3
5	The proposal assessment procedure is explicit, public and predefined and establishes rating and weighting of both technical and economic aspects.	T. 6	2
	Requirements for tender publication (amounts, time limits, publication media) established in legal or internal regulations		

6	(whichever is more stringent) are complied with.	T. 5	3
7	Framework contracts or agreements for direct supply of frequently used goods and services are in place.	T. 6	1
8	Quantitative and qualitative assessment of supplier and contractor performance, based on the results of indicators defined by the utility, and profile classification and updating, are conducted at least once a year	T. 4	1
9	The decision to renew or terminate contracts for outsourced services or to acquire frequently used goods and services is based on prior review of compliance with the conditions established in these contracts.	T. 4	1
10	The degree of internal users' satisfaction with the services provided by the procurement area is assessed.	T. 4	1

ME5.2 Purchases made by "public tender"

Definition: Percentage of purchases made by "public tender" as a proportion of all purchases made in the calendar year preceding the rating date, expressed in monetary terms.
Type: Indicator
Service: Drinking Water and/or Sanitation
Glossary: Public tender
Formula: ([EG5-V1]/[EG5-V2])*100   Unit: %
Normalization Function:

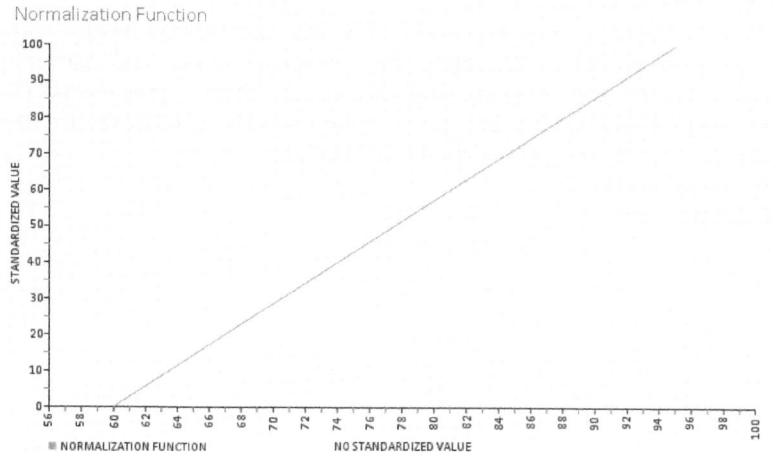

Variables

[EG5-V1] Value of purchases made by "public tender"
Definition: Value of purchases of goods and services made in the calendar year preceding the rating date by "public tender" awarded in the same year, or in previous years, including design and feasibility studies and works. Only the amount purchased in the year preceding the rating date will be considered, not the amount awarded.
Units: local currency
Reliability: Table 75

[EG5-V2] Value of total purchases
Definition: Value of total purchases of goods and services made in the calendar year preceding the rating date, including design and feasibility studies and works and excluding purchases made without "public tender" due to the fact that only one supplier exists for the respective good or service (this condition needs to be supported by documentation).
Units: local currency
Reliability: Table 76

ME5.3 "Successful tenders"

Definition: Percentage of "successful public" invitations to tender as a proportion of total "public" invitations to tender issued in the 3 calendar years preceding the rating date. If no public invitations to tender were issued in this period, data for the last year in which one was issued should be used.
Type: Indicator
Service: Drinking Water and/or Sanitation
Glossary: Public tender, Successful tenders
Formula: ([EG5-V3]/[EG5-V4])*100   Unit: %
Normalization Function:

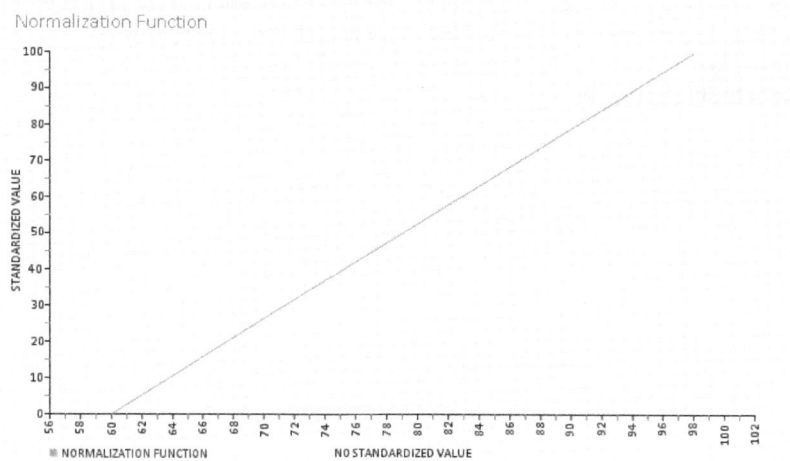

Variables

[EG5-V3] Number of "public tenders" awarded receiving 3 or more offers
Definition: Number of "public tenders" awarded receiving 3 or more offers of those held in the 3 calendar years preceding the rating date. If no public invitations to tender were issued in this period, data for the last year in which one was issued should be used.
Units: no.
Reliability: Table 77

[EG5-V4] Total "public" invitations to tender issued
Definition: Total "public" invitations to tender issued in the 3 calendar years preceding the rating date. If no public invitations to tender were issued in this period, data for the last year in which one was issued should be used.
Units: no.
Reliability: Table 78

ME5.4 Tenders held within the regulated "minimum timeframe"

Definition: Percentage of tenders held and awarded within the regulated "minimum timeframe", or which do not exceed it by more than 5%, as a proportion of total invitations to tender in the 3 full calendar years preceding the rating date. The timeframe starts on the date of the invitation to tender and ends on the date on which the awarding document is issued.
Type: Indicator
Service: Drinking Water and/or Sanitation
Glossary: Minimum tender period
Formula: ([EG5-V5]/[EG5-V6])*100   Unit: %
Normalization Function:

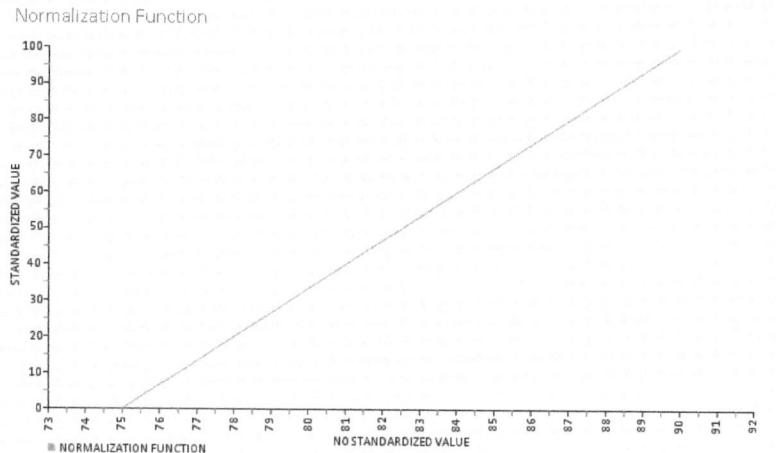

Variables

[**EG5-V5**] Number of tenders awarded
Definition: Number of tenders awarded of those issued in the 3 full calendar years preceding the rating date and which do not exceed the regulated "minimum tender period" by more than 5%.
Units: no.
Reliability: Table 79

[**EG5-V6**] Total invitations to tender issued
Definition: Total invitations to tender issued in the 3 full calendar years preceding the rating date.
Units: no.
Reliability: Table 80

ME6 Staff and support resource efficiency

The objective of this criterion is to assess utility management efficiency through staff productivity and the proportion of revenue allocated to support functions.

Indicators

ME6.1   Staff productivity
ME6.2   Expenditure on management and sales functions

ME6.1 Staff productivity

Definition: Ratio between the total number of "own staff" and the total number of "active drinking water and wastewater connections". The average data for the full calendar year preceding the rating date is considered in calculation.
Type: Indicator
Service: Drinking Water and/or Sanitation
Glossary: Own staff, Active connections
Formula: [EG4-V4]/([EG6-V2]/1000)   Unit: -
Normalization Function:

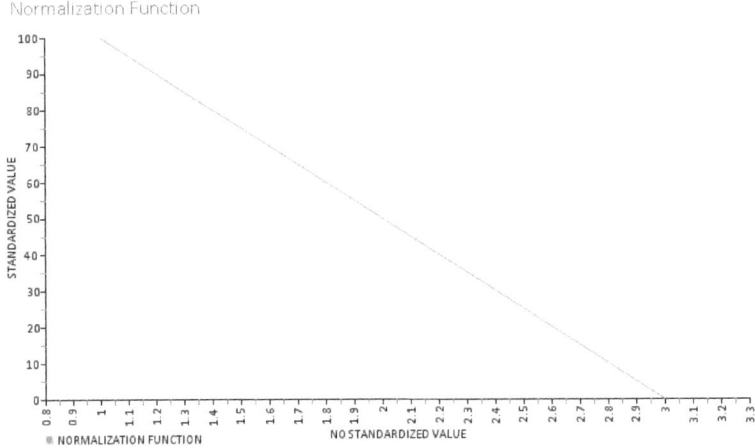

Variables

[EG4-V4] Total staff
Definition: Average total of "own staff" for the full calendar year preceding the rating date.
Units: no.
Reliability: Table 71

[EG6-V2] "Active" drinking water and wastewater connections
Definition: "Active" drinking water and wastewater connections (average for the preceding full calendar year).
Units: no.
Reliability: Table 81

ME6.2 Expenditure on management and sales functions

Definition: Ratio between expenditure on management and sales functions and revenue from delivery of drinking water and wastewater services. The average for the last 3 full calendar years is considered.
Type: Indicator
Service: Drinking Water and/or Sanitation
Glossary:
Formula: ([EG6-V3]/[SF3-V12])*100   Unit: %
Normalization Function:

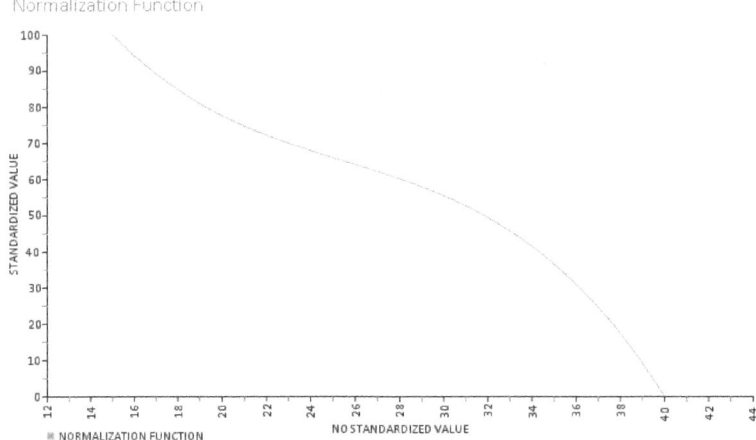

Variables

[EG6-V3] Expenses incurred in activities not directly related to service operation

Definition: Expenses incurred in activities not directly related to service operation, such as senior management, planning, legal advice, accounting, finance, human resources, information technology, procurement, customer service, billing, and collection. Expenses include benefits to staff who perform these activities, materials, services, depreciation and amortization, and other expenses incurred in performing these activities.

Units: financial statement currency

Reliability: Table 82

[SF3-V12] Revenue from service delivery

Definition: Revenue from service delivery or sales entered in the income statement or profit and loss statement for the period and consistent with the revenue entered in accounts receivable.

Units: financial statement currency

Reliability: Table 88

Note: For utilities for whom the scope of the Financial Statements does not coincide with the rating scope, a modified reliability table applies. See Table 1088

FS Financial Sustainability

The objective of this rating area is to assess the utility's capacity to finance business continuance based on the flows generated by service provision. It emphasizes the idea of not depending on State funds to cover total long-term costs, including return on capital.

This area includes three assessment sub-areas (overall financial sustainability, financial management and customer management) composed of practices and indicators. These three sub-areas enable comprehensive assessment of the utility's financial sustainability, since the first one measures the extent to which revenue from service delivery covers operating and finance expenses; the second one assesses the ability to meet financial obligations imposed by service delivery and to cover financial risk; and the third one assesses billing and collection management corresponding to service delivery.

Indicators are assessed on the basis of the average value for the last 3 calendar years preceding the rating date, unless measurement specifications indicate otherwise. Depending on the type of indicator, data for assessment are obtained either from the financial statements or from the customer information system. In the case of practices, those in force at the rating date or at the nearest date preceding that, as appropriate, are considered.

The sub-areas of evaluation are:

> FS1 Overall financial sustainability
> FS2 Financial management
> FS3 Customer management

FS1 Overall financial sustainability

The objective of this sub-area is to assess to what extent revenue from service delivery consistently covers total costs in order to ensure financial sustainability. Analysis considers a range of degrees, starting from when this revenue is insufficient to cover disbursable operating expenses, up to the point of generating a surplus that enables payment of remuneration to capital (return on equity). Additionally, it includes assessment of the practices related to service tariffs and to basic information necessary to monitor and manage the elements that have an impact on financial sustainability, considering as basic information financial statements, financial projections and cost information.

Practices

    FS1.1   Financial sustainability
    FS1.2   Expense coverage

Indicators

    FS1.3   "Return on equity"

FS1.1 Financial sustainability

Type: Best Practices
Service: Drinking Water and/or Sanitation
Normalization: Weighted by practices
Glossary: System, Total long-term costs, Appropriate cost of capital, Costs per activity, Service stage
Definition: Includes:

Practices	Reliability	Weight
1	Tariffs are calculated to cover the "total long-term costs" of service delivery for a period not shorter than 15 years. These tariffs are determined on the basis of a study completed within the last 5 calendar years preceding the rating period and considering an "adequate rate of capital cost".	T. 83	10
2	An automatic tariff indexation mechanism exists and is applied. (This practice is considered complied with at maximum reliability level when in the respective country inflation in the last 3 years has not exceeded 3% per annum).	T. 84	5
3	The tariff structure is differentiated by service type (drinking water supply, wastewater collection, wastewater treatment), "system", geographical area or service area, as appropriate. (This practice is considered complied with at maximum reliability level when the utility operates just one "system" and provides just one of the services).	T. 85	2
4	In the case that a subsidy to pay for service and/or to connect to the drinking water and/or wastewater network is granted, the subsidy's source of finance is known, stable and sufficient, irrespective of whether the source is external or internal (cross-subsidy). (This practice is considered complied with at maximum reliability level if both practices AS1.1.4 and AS1.1.5 are declared not complied with, or if practice AS1.1.4 is declared not complied with and practice AS1.1.5 is considered complied with due to fulfillment of the service coverage criteria there defined).	T. 6	3
	Complete annual financial statements are produced and prepared according to international accounting principles and standards (IFRS) and are reviewed by external auditors,		

5	whose report is issued within the first 3 months of the following year.	T. 86	2
6	Complete monthly financial statements are produced and prepared according to international accounting standards (IFRS) or to the standards in effect in the country.	T. 86	3
7	Up-to-date financial projections for at least the next 5 years are available. These projections include the balance sheet, income statement and cash flow statement.	T. 87	2
8	A cost system exists that delivers information by cost center, "service stage" (drinking water production, drinking water distribution, wastewater collection, wastewater treatment and discharge) and "activity".	T. 3	1
9	The accounting system provides reports on revenue, expenditure and income by business unit/locality and by type of service delivered.	T. 3	2

FS1.2 Expense coverage

This assessment element assesses if revenue from delivery of drinking water and/or wastewater services, depending on the service(s) being rated, is sufficient to cover "operating" and "financial" expenses. Four levels of expense coverage are considered in ascending order. All cost coverage levels complied with must be indicated. Consequently, if a higher level is complied with, all levels below it must also be indicated since they are also complied with (levels are not mutually exclusive). The four levels are as follows:

Type: Best Practices
Service: Drinking Water and/or Sanitation
Normalization: Weighted by practices
Glossary: Operational expenses, Financial expenses
Definition: Degree to which revenue from delivery of drinking water and/or wastewater services, depending on the service(s) being rated, covers "operating" and "financial" expenses in the rating period.
Note: For utilities for whom the scope of the Financial Statements does not coincide with the rating scope, a modified reliability table applies. See Table 1088

Practices		Reliability	Weight
1	Revenue from delivering drinking water and/or wastewater services covers "operating expenses", excluding depreciation and amortization.	T. 88	3
2	Revenue from delivering drinking water and/or wastewater services covers "operating expenses", excluding depreciation and amortization, and also covers "financial expenses".	T. 88	2
3	Revenue from delivering drinking water and/or wastewater services covers "operating expenses", including depreciation and amortization, but does not cover "financial expenses".	T. 88	3
4	Revenue from delivering drinking water and/or wastewater services covers "operating expenses", including depreciation and amortization, and also covers "financial expenses".	T. 88	2

FS1.3 "Return on equity"

Definition: Measures the return on resources invested by the "owners" in the rating period.
Type: Indicator
Service: Drinking Water and/or Sanitation
Glossary: Owner (utility), Return on equity
Formula: $((([SF1-V1]/[SF1-V2])-[SF1-V3])/(1+[SF1-V3]))*100$ Unit: %
Normalization Function:

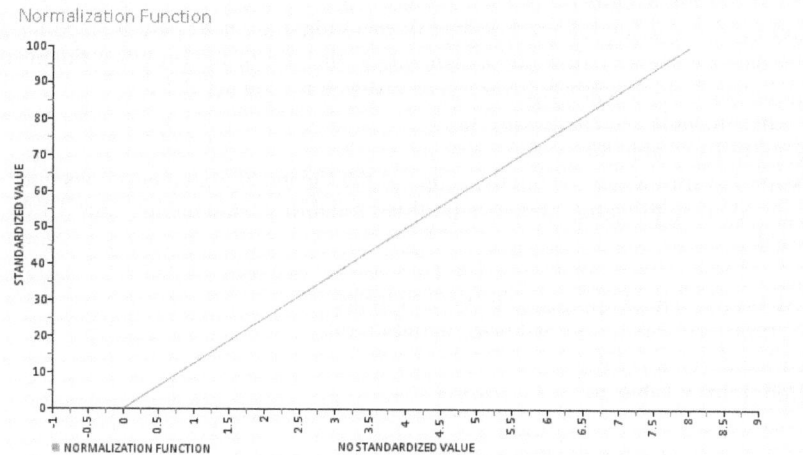

Variables

[SF1-V1] Income for the financial period
Definition: Net income for the financial period after tax.
Units: financial statement currency
Reliability: Table 88
Note: For utilities for whom the scope of the Financial Statements does not coincide with the rating scope, a modified reliability table applies. See Table 1088

[SF1-V2] Initial equity
Definition: Amount entered under this denomination or as paid-in capital on the balance sheet or financial statements at the end of the previous financial period and which represents resources invested by "owners".
Units: financial statement currency
Reliability: Table 88
Note: For utilities for whom the scope of the Financial Statements does not coincide with the rating scope, a modified reliability table applies. See Table 1088

[SF1-V3] Annual internal price variation rate.
Definition: Annual internal price variation rate.
Units: parts per unit
Reliability: Table 56

FS2 Financial management

This section's objective is to assess the entity's liquidity and financial solvency, i.e. its capacity to access financing, to meet its short- and long-term financial obligations, and to hedge financial risk. In addition, application of mechanisms to assess and improve the internal control system is rated due to its impact on appropriate management of financial resources.

Practices

FS2.1   Financing, risk hedging and internal control

Indicators

FS2.2   Current ratio
FS2.3   Debt-to-equity ratio
FS2.4   Committed flows
FS2.5   Currency risk
FS2.6   Rate risk

FS2.1 Financing, risk hedging and internal control

Type: Best Practices
Service: Drinking Water and/or Sanitation
Normalization: Weighted by practices
Glossary: Public debt instruments, In force (investment plan in force)
Definition: Includes:

Practices	Reliability	Weight
1 More than 50% of external financing for the "current" investment program comes from independent institutions (multilateral banks, international cooperation agencies, publicly owned financial institutions or privately owned financial institutions) and/or from issuance of "public debt instruments" by the utility.	T. 89	1
2 More than 50% of external financing for the "current" investment program comes from privately owned financial institutions and/or issuance of "public debt instruments" by the utility.	T. 89	1
3 A financial risk management policy exists based on financial risk analysis. This considers use of hedging instruments to mitigate interest rate and foreign currency risk. (This practice is considered complied with at maximum reliability level if the utility has not entered financial liabilities in its financial statements in the last 2 calendar years preceding the rating date).	T. 2	1
4 The internal audit unit prepares an annual work program based on analysis of internal control risks.	T. 2	1
5 A duly approved internal audit handbook or rules of practice is available.	T. 90	1
6 The internal audit unit issues reports on reviews. The observations are followed up and evidence exists that the issues have been addressed, or justifications are given when an issue has not been addressed.	T. 4	2

External auditors, or the entity responsible for external financial control of the utility, review the internal control

7 | system annually and evidence exists that the issues indicated in their report have been addressed, or if an issue has not been addressed, this action is justified. | T. 4 | 2

FS2.2 Current ratio

Definition: Capacity to pay debt due within a year of the closing date of the financial period. It is expressed in number of times. The average for the last 3 full calendar years is considered.
Type: Indicator
Service: Drinking Water and/or Sanitation
Glossary:
Formula: ([SF2-V1]/[SF2-V2])   Unit: Number of times
Normalization Function:

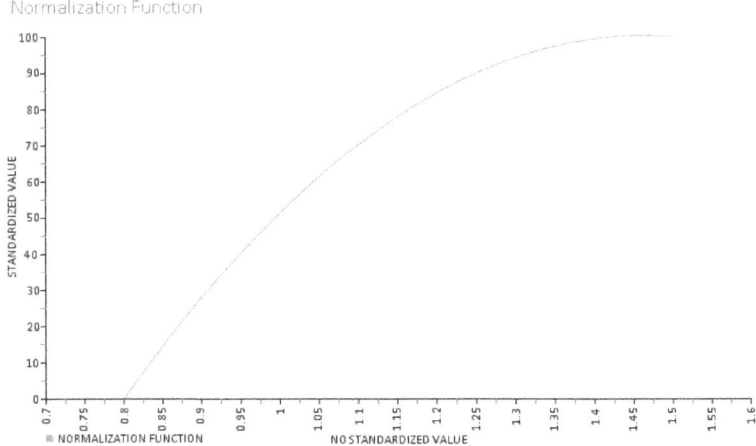

Variables

[SF2-V1] Current assets
Definition: Amount entered under this denomination or as liquid assets on the balance sheet or financial statements. Current assets comprise cash and equivalents to cash at the closing date of the financial period and of goods and rights susceptible to consumption or conversion into money during the next 12 months.
Units: financial statement currency
Reliability: Table 88
Note: For utilities for whom the scope of the Financial Statements does not coincide with the rating scope, a modified reliability table applies. See Table 1088

[SF2-V2] Current liabilities
Definition: Amount entered under this denomination on the balance sheet or financial statements. Current liabilities comprise commitments and obligations to third parties due for payment within the 12 months following the closing date of the financial period.
Units: financial statement currency
Reliability: Table 88
Note: For utilities for whom the scope of the Financial Statements does not coincide with the rating scope, a modified reliability table applies. See Table 1088

FS2.3 Debt-to-equity ratio

Definition: Measures the financing structure in use by calculating the ratio between short- and long-term liabilities and equity. It is also known as leverage. It is expressed in number of times. The average for the last 3 full calendar years is considered.
Type: Indicator
Service: Drinking Water and/or Sanitation
Glossary:
Formula: ([SF2-V5]/[SF2-V4])   Unit: Number of times
Normalization Function:

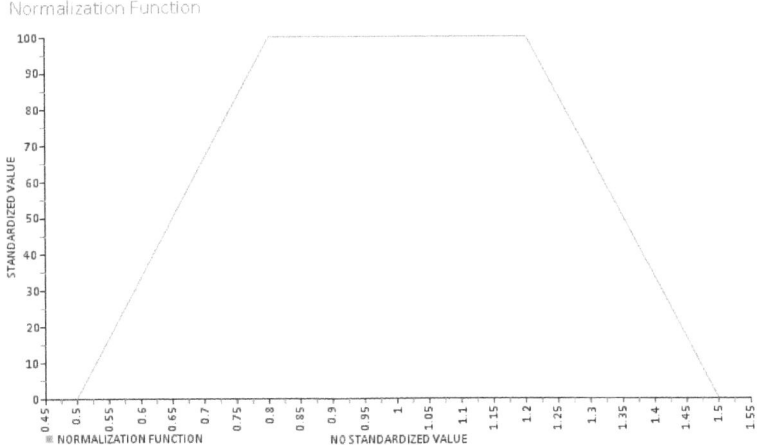

Variables

[SF2-V4] Net worth
Definition: Amount entered under this denomination on the balance sheet or financial statements at the end of the financial period.
Units: financial statement currency
Reliability: Table 88
Note: For utilities for whom the scope of the Financial Statements does not coincide with the rating scope, a modified reliability table applies. See Table 1088

[SF2-V5] Total liabilities
Definition: Sum of current and non-current liabilities entered on the balance sheet or financial statements. They represent all obligations and commitments to third parties at the end of the financial period.
Units: financial statement currency
Reliability: Table 88
Note: For utilities for whom the scope of the Financial Statements does not coincide with the rating scope, a modified reliability table applies. See Table 1088

FS2.4 Committed flows

Definition: Measures the number of years in terms of cash flow represented by total liabilities. The average for the last 3 full calendar years is considered.
Type: Indicator
Service: Drinking Water and/or Sanitation
Glossary:
Formula: [SF2-V5]/([SF2-V3])   Unit: Number of years
Normalization Function:

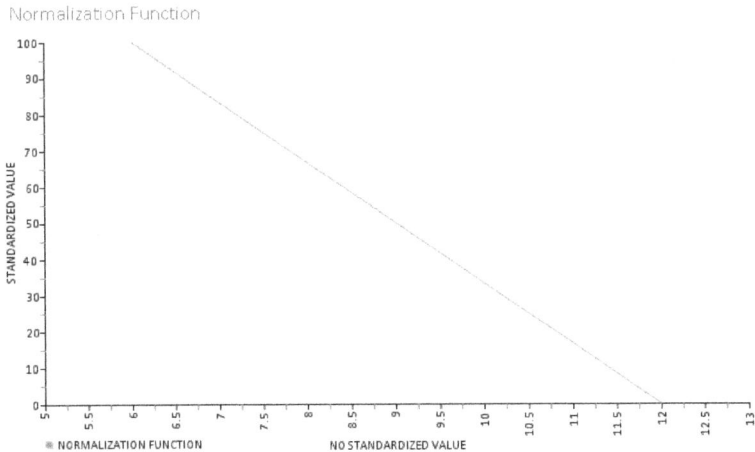

Variables

[SF2-V3] EBITDA
Definition: Represents an approximation to cash flow derived from operations. It is computed from the income statement or profit and loss statement for the financial period and may be obtained in two ways, depending on how data are presented in this statement: i) If data are classified by nature, EBITDA = revenue from regular activities or operations minus costs of raw materials and consumables, employee benefits and other expenses, or ii) if data are classified by function, EBITDA = earnings before interest and tax plus depreciation and amortization (EBIT).
Units: financial statement currency
Reliability: Table 88
Note: For utilities for whom the scope of the Financial Statements does not coincide with the rating scope, a modified reliability table applies. See Table 1088

[SF2-V5] Total liabilities
Definition: Sum of current and non-current liabilities entered on the balance sheet or financial statements. They represent all obligations and commitments to third parties at the end of the financial period.
Units: financial statement currency
Reliability: Table 88
Note: For utilities for whom the scope of the Financial Statements does not coincide with the rating scope, a modified reliability table applies. See Table 1088

FS2.5 Currency risk

Definition: Measures the proportion of debt owed in an indexed currency or in a currency different to the one in which sales revenue is earned, without hedging, as a proportion of total debt. The average for the last 3 full calendar years is considered.
Type: Indicator
Service: Drinking Water and/or Sanitation
Glossary:
Formula: ([SF2-V8]/[SF2-V5])*100   Unit: %
Normalization Function:

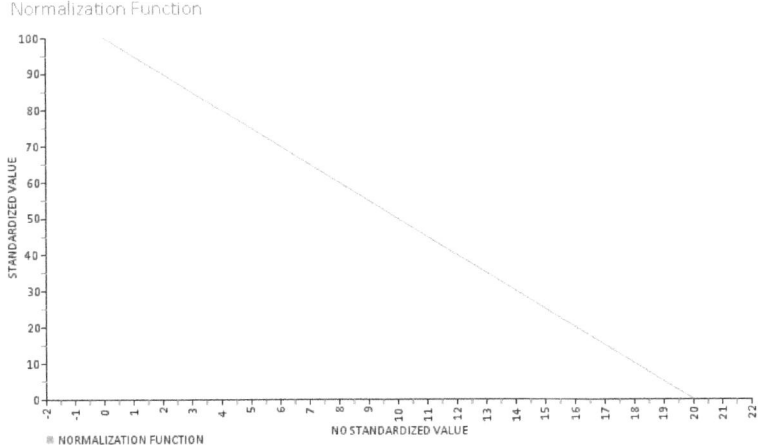

Variables

[SF2-V5] Total liabilities
Definition: Sum of current and non-current liabilities entered on the balance sheet or financial statements. They represent all obligations and commitments to third parties at the end of the financial period.
Units: financial statement currency
Reliability: Table 88
Note: For utilities for whom the scope of the Financial Statements does not coincide with the rating scope, a modified reliability table applies. See Table 1088

[SF2-V8] Liabilities in foreign or indexed currency without hedging
Definition: Amount of debt owed in an indexed currency or in a currency other than the one in which sales revenue is earned, discounting the hedged amount, at the closing date of the financial period.
Units: financial statement currency
Reliability: Table 88
Note: For utilities for whom the scope of the Financial Statements does not coincide with the rating scope, a modified reliability table applies. See Table 1088

FS2.6 Rate risk

Definition: Measures the proportion of debt owed at a variable rate, without hedging, as a proportion of financial debt. The average for the last 3 full calendar years is considered.
Type: Indicator
Service: Drinking Water and/or Sanitation
Glossary:
Formula: ([SF2-V9]/[SF2-V10])*100   Unit: %
Normalization Function:

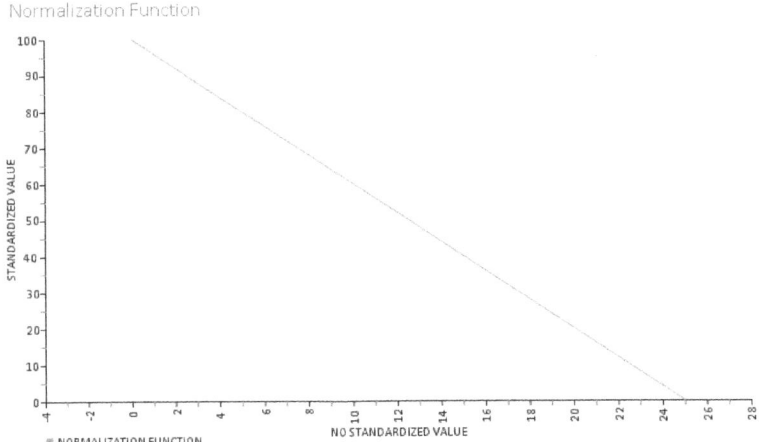

Variables

[SF2-V10] Financial liabilities
Definition: Amount of short- and long-term financial debt entered on the balance sheet or financial statements under current liabilities and non-current liabilities at the closing date of the financial period.
Units: financial statement currency
Reliability: Table 88
Note: For utilities for whom the scope of the Financial Statements does not coincide with the rating scope, a modified reliability table applies. See Table 1088

[SF2-V9] Liabilities without hedging owed at a variable rate
Definition: Amount of debt owed at a variable interest rate, discounting the hedged amount, at the closing date of the financial period.
Units: financial statement currency
Reliability: Table 88
Note: For utilities for whom the scope of the Financial Statements does not coincide with the rating scope, a modified reliability table applies. See Table 1088

FS3 Customer management

Considering the importance of appropriate revenue management to the financial sustainability of drinking water and wastewater utilities, this criterion assesses management and practices related to billing, collection and recovery of revenue for these services.

Practices

FS3.1   Billing and collection

Indicators

FS3.2   Billing effectiveness
FS3.3   Billing error rate
FS3.4   Unbilled water
FS3.5   Collection rate
FS3.6   Average collection time
FS3.7   Arrearage

FS3.1 Billing and collection

Type: Best Practices
Service: Drinking Water and/or Sanitation
Normalization: Weighted by practices
Glossary: Property
Definition: Includes:

Practices	Reliability	Weight
1 Billing is based on measurement (meter reading) for at least 99% of billed users.	T. 93	3
2 The user register and classification is updated within 10 days of establishment of a service agreement or of deactivation of a connection.	T. 94	1
3 A user register exists that includes information concerning user type, service status (active/inactive), meter data, "property" data and other information necessary for billing the service.	T. 33	1
4 Data scanners with a magnetic interface to the billing system or remote meter-reading technologies are in use.	T. 95	2
5 The quality of meter reading and bill issue in each zone or sector is systematically monitored.	T. 95	2
6 Services are billed monthly or bimonthly or payments are received with this regularity if bills are issued less frequently.	T. 93	1
7 The bill format contains all data required to verify computation of the amount billed, comprising at the very least the following: amount payable and due date (highlighted), reading date and quantities (current and previous), volume consumed, volume billed and tariff for each service billed, and other charges or adjustments applied.	T. 96	1
8 Formally defined collection procedures exist that include systematic use of coercive recovery instruments for users in arrears, to the extent allowed by legislation.	T. 93	2

9   A policy exists for detecting and regularizing fraud in its various forms (meter calibration, detection of illegal connections, detection of false information regarding type of use or any other variable that influences tariffs) and, in the case that estimated losses attributable to users exceed 10% of unbilled water volume, systematic operations are carried out to detect illegal connections. If an estimate of losses attributable to users is not available, it is assumed that these amount to more than 10% of unbilled water volume.   T. 97   2

FS3.2 Billing effectiveness

Definition: Number of service users billed as a proportion of total "active users". The average rate for the calendar year preceding the rating date is considered.
Type: Indicator
Service: Drinking Water and/or Sanitation
Glossary: Registered user, Active users, Zero-value bill
Formula: ([SF3-V1]/[SF3-V2])*100   Unit: %
Normalization Function:

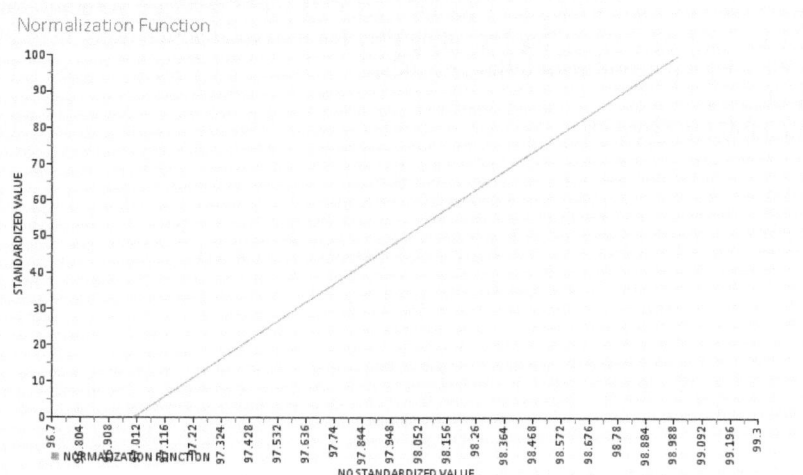

Variables

[SF3-V1] Number of users billed, excluding "zero-value bills"
Definition: Number of users billed, excluding "zero-value bills" Average for the calendar year preceding the rating date (sum of number of users billed in each billing period [excluding zero-value bills] / number of billing periods).
Units: no.
Reliability: Table 98

[SF3-V2] Total "active users"
Definition: "Users registered" with the utility to use or consume drinking water and/or wastewater services and receiving those services. Average for the calendar year preceding the rating date (sum of total active users in each billing period / number of billing periods).
Units: no.
Reliability: Table 98

FS3.3 Billing error rate

Definition: Number of bills reissued or amended due to errors detected cither by internal controls or by user "complaints" as a proportion of total bills issued. The average rate for the calendar year preceding the rating date is considered.

Type: Indicator
Service: Drinking Water and/or Sanitation
Glossary: Complaint
Formula: ([SF3-V3]/[SF3-V4])*100   Unit: %
Normalization Function:

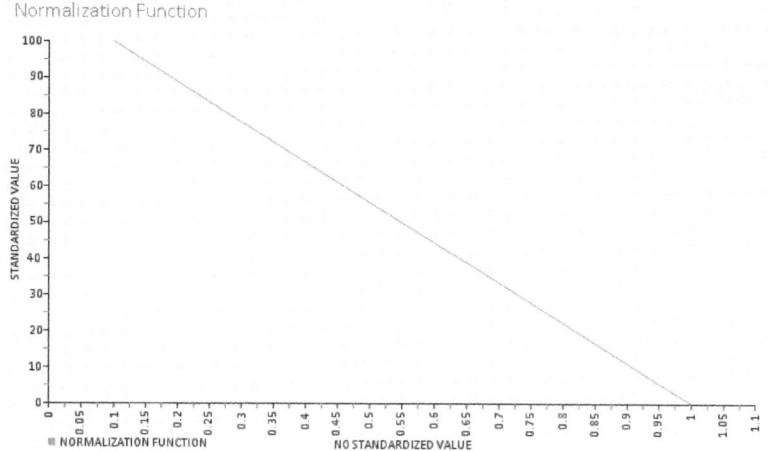

Variables

[**SF3-V3**] Number of bills reissued or amended
Definition: Number of bills reissued or amended due to errors detected either by internal controls or by user "complaints". Average for the calendar year preceding the rating date (sum of number of bills reissued or amended in each billing period / number of billing periods).
Units: no.
Reliability: Table 98

[**SF3-V4**] Total number of bills issued
Definition: Total number of bills issued Average for the calendar year preceding the rating date (sum of number of bills issued in each billing period / number of billing periods).
Units: no.
Reliability: Table 98

FS3.4 Unbilled water

Definition: Proportion of water introduced into the "system" that is not billed in the rating period.

Type: Indicator

Service: Drinking Water and/or Sanitation

Glossary: System, Water volume incorporated into the system

Formula: If connections density < 20

    $(([EO1-V2]-[SF3-V11])/[EO1-V2])*100$ Unit: %

If connections density >= 20

    $(([EO1-V2]-[SF3-V11])/[EO1-V2])*100$ Unit: %

Normalization Function:

If connections density < 20

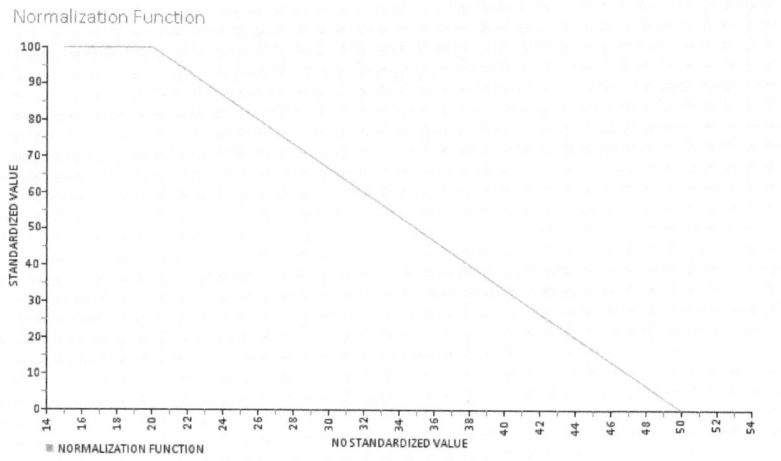

FS Financial Sustainability

If connections density >= 20

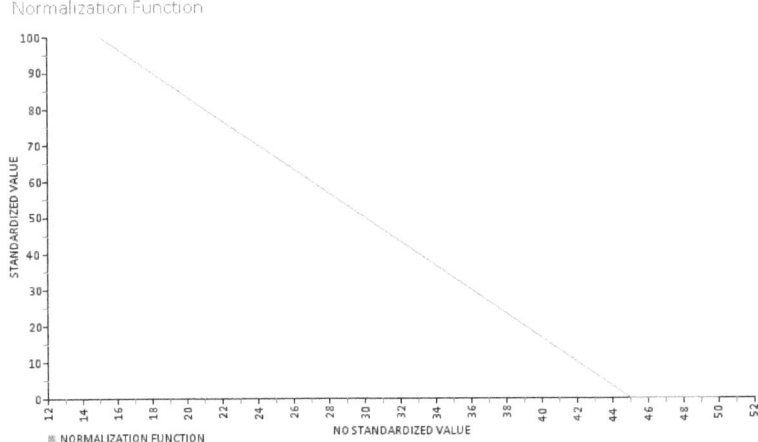

Variables

[EO1-V2] Total "water volume introduced into the system"
Definition: Total "water volume introduced into the system"
Units: m3
Reliability: Table 41

[SF3-V11] Billed water volume
Definition: Total billed water volume in the rating period, according to billing system records.
Units: m3
Reliability: Table 98

FS3.5 Collection rate

Definition: Proportion of collectible revenue for services billed collected in the rating period.
Type: Indicator
Service: Drinking Water and/or Sanitation
Glossary:
Formula: ([SF3-V5]/[SF3-V6])*100   Unit: %
Normalization Function:

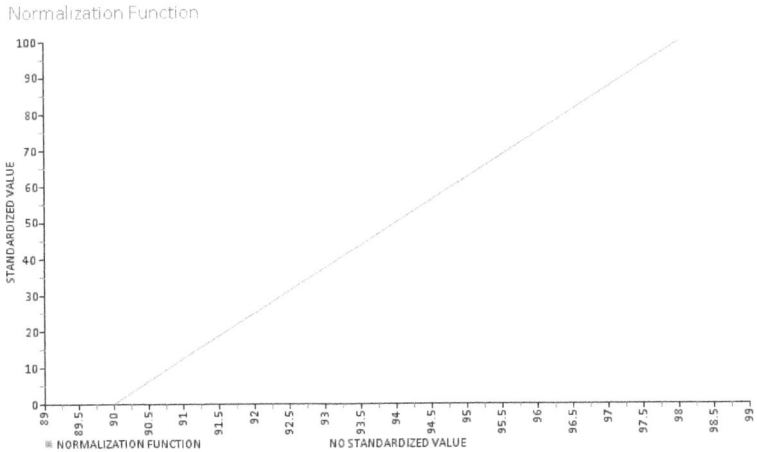

Variables

[**SF3-V5**] Collections for services delivered
Definition: Amount collected in the rating period for services delivered.
Units: financial statement currency
Reliability: Table 99

[**SF3-V6**] Revenue collectible for services delivered
Definition: Amount of revenue for services delivered billed in the rating period, including sales tax if applicable, minus accounts receivable due to mature by the closing date of the financial period, plus the balance of accounts receivable due to mature at the end of the previous financial period and those overdue by up to 90 days.
Units: financial statement currency
Reliability: Table 99

FS3.6 Average collection time

Definition: Average time taken to collect payment for services delivered, expressed in number of days. The average for the last 3 full calendar years is considered.
Type: Indicator
Service: Drinking Water and/or Sanitation
Glossary: Weighted average tax rate
Formula: [SF3-V7]/([SF3-V12]*(1+[SF3-V8])/360)   Unit: days
Normalization Function:

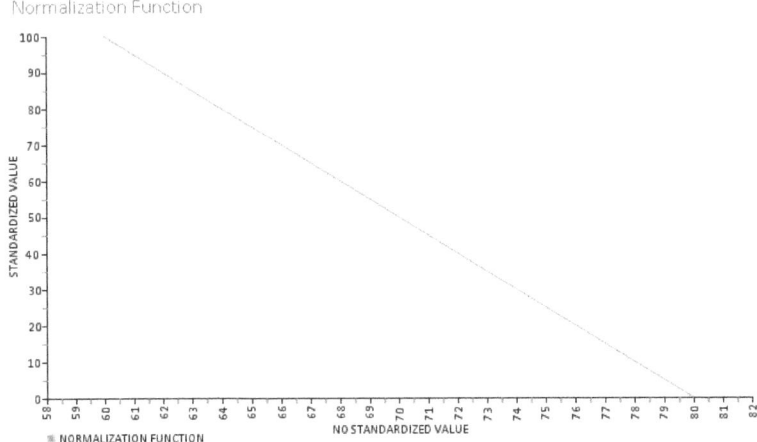

Variables

[SF3-V12] Revenue from service delivery
Definition: Revenue from service delivery or sales entered in the income statement or profit and loss statement for the period and consistent with the revenue entered in accounts receivable.
Units: financial statement currency
Reliability: Table 88
Note: For utilities for whom the scope of the Financial Statements does not coincide with the rating scope, a modified reliability table applies. See Table 1088

[SF3-V7] Balance of accounts receivable (gross value without discounting the allowance for uncollectible receivables)
Definition: Amount of accounts receivable for services delivered at the end of the financial period, without discounting allowance for uncollectible receivables.
Units: financial statement currency
Reliability: Table 88
Note: For utilities for whom the scope of the Financial Statements does not coincide with the rating scope, a modified reliability table applies. See Table 1088

[SF3-V8] Sales tax rate
Definition: Sales tax rate or value-added tax (VAT) rate. In the case that different tax rates are applied to sales, or that one portion of sales is taxable while another is exempted from taxation, the "weighted average tax rate" will be calculated according to the percentage represented by total revenue for service delivery subject to the various tax rates and, if applicable, by revenue exempt from tax, within total revenue for services delivered (variable FS3-V12) entered in the financial statements. See glossary for weighted average tax rate calculation formula.
Units: -
Reliability: Table 56

FS3.7 Arrearage

Definition: Proportion of accounts receivable not paid within 90 days. The average for the last 3 full calendar years is considered.
Type: Indicator
Service: Drinking Water and/or Sanitation
Glossary:
Formula: ([SF3-V10]/[SF3-V7])*100   Unit: %
Normalization Function:

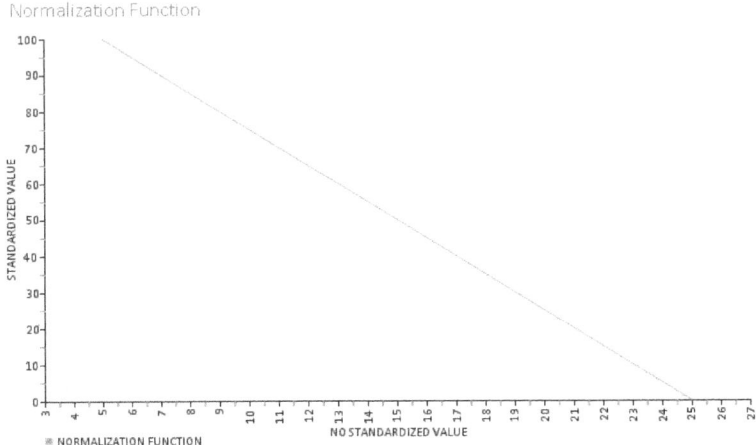

Variables

[SF3-V10] Accounts unpaid for more than 90 days
Definition: Accounts unpaid for more than 90 days at the closing date of the financial period
Units: financial statement currency
Reliability: Table 99

[SF3-V7] Balance of accounts receivable (gross value without discounting the allowance for uncollectible receivables)
Definition: Amount of accounts receivable for services delivered at the end of the financial period, without discounting allowance for uncollectible receivables.
Units: financial statement currency
Reliability: Table 88

AS Access to Service

Access to drinking water and wastewater services is fundamental to household health and quality of life, as well as to economic and social development. The United Nations explicitly recognizes access to these services as a human right. Therefore, drinking water and wastewater service coverage is considered an essential area in this rating system, which complements the service quality assessment presented in the SQ area.

Assessment centers on services delivered via network infrastructure, although it also considers cases in which the utility is obliged to deliver services temporarily by other means, such as tankers. This area assesses practices followed to facilitate service access to the entire population within the geographical area to be rated, as well as assessing a series of quantitative indicators that enable rating of the degree of physical service coverage and the financial accessibility of the service.

It should be noted that the utility is responsible for a wide variety of obligations which range from universal coverage within an entire administrative area (district, metropolitan area, etc.) to excluding from service delivery certain zones within that geographical area (for example neighborhoods that do not comply with minimum urban development standards). This circumstance is considered by explicitly defining physical coverage measures in relation to the geographical service coverage area defined in the utility's mandate.

The Access to Service area is not divided into sub-areas.

Practices

AS1.1   Guaranteed "access" to service

Indicators

AS1.2   Household "access" to drinking water
AS1.3   Connection to wastewater collection "systems"
AS1.4   Household ability to pay for services received

AS1.1 Guaranteed "access" to service

Type: Best Practices
Service: Drinking Water and/or Sanitation
Normalization: Weighted by practices
Glossary: System, Access, Geographical area to be rated, Competent official body
Definition: Includes:

Practices	Reliability	Weight
1 Plans exist to extend the drinking water service to households currently without household "access". The plans have coverage goals and established timeframes that are equal to or more stringent than those set by the pertinent authority. (This practice is considered complied with at maximum reliability level if the proportion of the population with a household connection to the drinking water network in the "geographical area to be rated" is higher than 99.9% for the year preceding the rating date, provided that figure is supported by data published by a "competent official body" or is supplied by indicator AS1.2 (>99.9%) and average reliability for that indicator stands at 0.8 or more).	T. 2	3
2 Alternative drinking water supply programs exist for zones that do not have household connections in the "geographical area to be rated" for drinking water supply, assuring service availability at distances of less than 500 m from each dwelling and assuring service continuity and quality. (This practice is considered complied with at maximum reliability level if the proportion of the population with a household connection to the drinking water network in the "geographical area to be rated" is higher than 99.9% for the year preceding the rating date, provided that figure is supported by data published by a "competent official body" or is supplied by indicator AS1.2 (>99.9%) and average reliability for that indicator stands at 0.8 or more).	T. 121	3

Plans exist to extend the wastewater collection service to households currently without household access to the mains "system". The plans have coverage goals and established timeframes that are equal to or more stringent than those set by the pertinent authority. (This practice is considered

3   complied with at maximum reliability level if the proportion of the population with a household connection to the wastewater collection network in the "geographical arca to be rated" is higher than 99.9% for the year preceding the rating date, provided that figure is supported by data published by a "competent official body" or is supplied by indicator AS1.3 (>99.9%) and average reliability for that indicator stands at 0.8 or more).   T. 2   3

4   A special tariff system or subsidy program exists for low-income households that facilitates payment for regular consumption of drinking water and/or wastewater services.   T. 2   1

5   A special tariff system or program exists for low-income households that facilitates payment for connection to the drinking water and/or wastewater networks, either through a subsidy or through a loan. (This practice is considered complied with at maximum reliability level if the proportion of the population with a household connection to both the drinking water and wastewater collection "systems" in the "geographical area to be rated" is higher than 99.9% for the year preceding the rating date, provided that figure is supported by data published by a "competent official body" or is supplied by indicators AS1.2 and AS1.3 (>99.9%) and average reliability for those indicators stands at 0.8 or more).   T. 2   3

6   A specific department or operations unit exists for planning and attending to zones without drinking water and/or wastewater service coverage. (This practice is considered complied with at maximum reliability level if the proportion of the population with a household connection to both the drinking water and wastewater collection "systems" in the "geographical area to be rated" is higher than 99.9% for the year preceding the rating date, provided that figure is supported by data published by a "competent official body" or is supplied by indicators AS1.2 and AS1.3 (>99.9%) and average reliability for those indicators stands at 0.8 or more).   T. 2   2

AS1.2 Household "access" to drinking water

This element considers the degree of coverage of the drinking water supply "system" (for use and consumption). Assessment is based on the proportion of the population located in the utility's "geographical area to be rated" receiving the drinking water supply service either in their dwellings or on their plots ("properties").

Other types of commercial or industrial activities are not considered, since they are not as clearly linked to the objective of this rating area as households.

Definition: Percentage of inhabitants with household "access" to drinking water via distribution networks as a proportion of total inhabitants in the utility's "geographical area to be rated" for drinking water supply in the calendar year preceding the rating date.
Type: Indicator
Service: Drinking Water
Glossary: System, Property, Access, Geographical area to be rated
Formula: ([AS-V1]/[AS-V2])*100   Unit: %
Normalization Function:

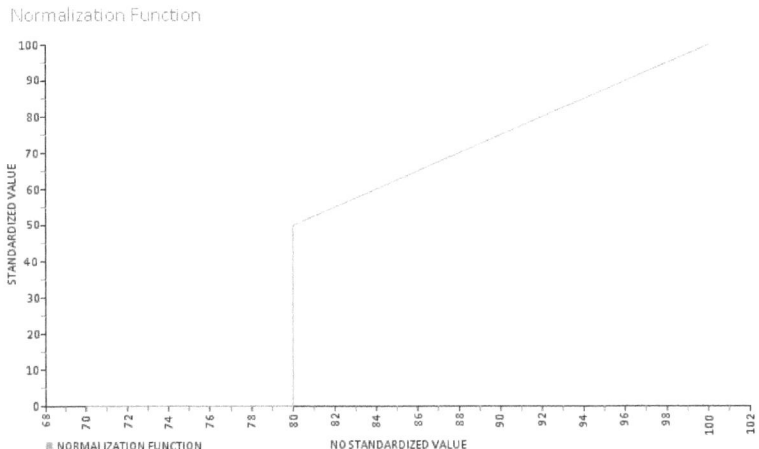

Variables

[AS-V1] Number of inhabitants with household "access" to the drinking water distribution network.
Definition: Number of inhabitants with household "access" to the drinking water distribution network in the "geographical area to be rated" for drinking water supply at the end of the calendar year preceding the rating date.
Units: no.
Reliability: Table 100

[AS-V2] Total inhabitants in the "geographical area to be rated" for drinking water supply
Definition: Total inhabitants in the "geographical area to be rated" for drinking water supply at the end of the calendar year preceding the rating date.
Units: no.
Reliability: Table 101

AS1.3 Connection to wastewater collection "systems"

This element considers the degree of coverage provided by the wastewater collection network. Assessment is based on the proportion of the population located in the "geographical area to be rated" that has a discharge point in its dwelling or on property with operative sewer infrastructure duly connected to the collection system (transporting the wastewater to a treatment plant or, in its absence, to a water body with drainage capacity or for integration into the aquatic environment). In cases in which the municipality or competent authority authorizes use of septic tanks, these will be considered equivalent to existence of a connection to a wastewater collection network.

Other types of commercial or industrial activities are not considered, since they are not as clearly linked to the objective of this rating area as households, even though they may have an impact on public health in the zone.

Service delivery quality is not considered either, except in accepting proper operation of the wastewater collection network, or septic tanks if applicable.

This assessment element considers only wastewater drainage capacity.

Definition: Number of inhabitants with a household connection to the wastewater collection network as a proportion of the total number of inhabitants in the "geographical area to be rated" for wastewater collection at the end of the calendar year preceding the rating date. (In cases in which the municipality or competent authority authorizes use of septic tanks in the "geographical area to be rated", these will be considered equivalent to existence of a connection to a wastewater collection network).
Type: Indicator
Service: Sanitation
Glossary: System, Geographical area to be rated
Formula: $([AS-V4]/[CS3-V4])*100$ Unit: %
Normalization Function:

AquaRating

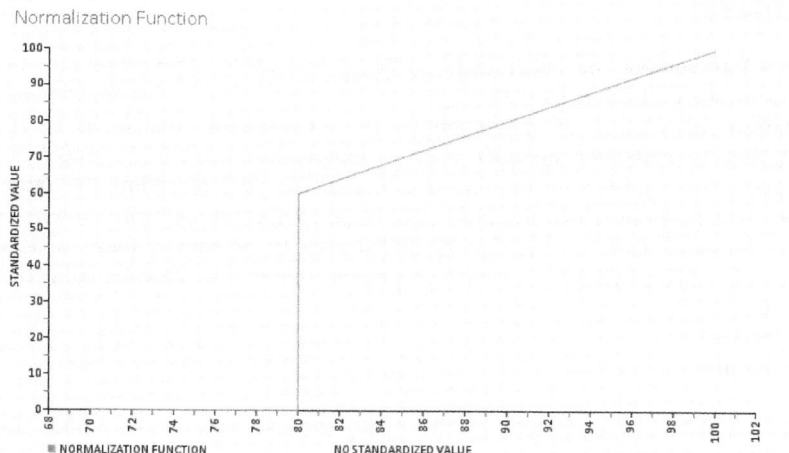

Normalization Function

Variables

[AS-V4] Number of inhabitants with a household connection to the wastewater collection network

Definition: Number of inhabitants with a household connection to the wastewater collection network in the "geographical area to be rated" for wastewater collection at the end of the calendar year preceding the rating date. (In cases in which the municipality or competent authority authorizes use of septic tanks in the "geographical area to be rated", these will be considered equivalent to existence of a connection to a wastewater collection network).

Units: no.

Reliability: Table 100

[CS3-V4] Total inhabitants in the "geographical area to be rated" for wastewater collection.

Definition: Total inhabitants in the "geographical area to be rated" for wastewater collection at the end of the calendar year preceding the rating date.

Units: inhabitants

Reliability: Table 101

AS1.4 Household ability to pay for services received

This element assesses the impact of expenditure on the drinking water service (and wastewater service if applicable, according to "rating scope") on family finances in low-income households, taking into consideration possible subsidies received by these households.

Definition: Monthly expenditure on the drinking water service (and wastewater service if applicable, according to "rating scope") per household, expressed as a percentage of average monthly household income in the poorest quintile. Expenditure is calculated by assuming consumption of 200 liters per household per day and applying the tariff for this volume of consumption/wastewater discharge.

Type: Indicator

Service: Drinking Water and/or Sanitation

Glossary: Geographical area to be rated, Rating scope

Formula: If the utility provides drinking water and sanitation services
$\quad$ ([AS-V5.1]/[AS-V6])*100 $\qquad$ Unit: %
If the utility provides drinking water services
$\quad$ ([AS-V5.2]/[AS-V6])*100 $\qquad$ Unit: %

Normalization Function:
$\quad$ **If the utility provides drinking water and sanitation services**

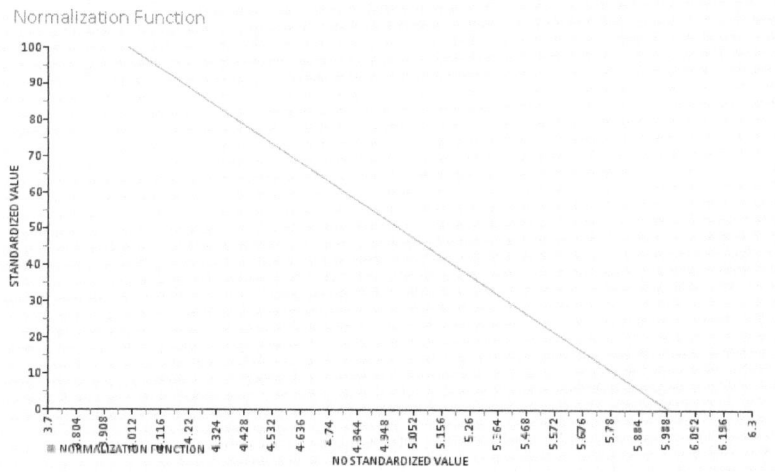

If the utility provides drinking water services

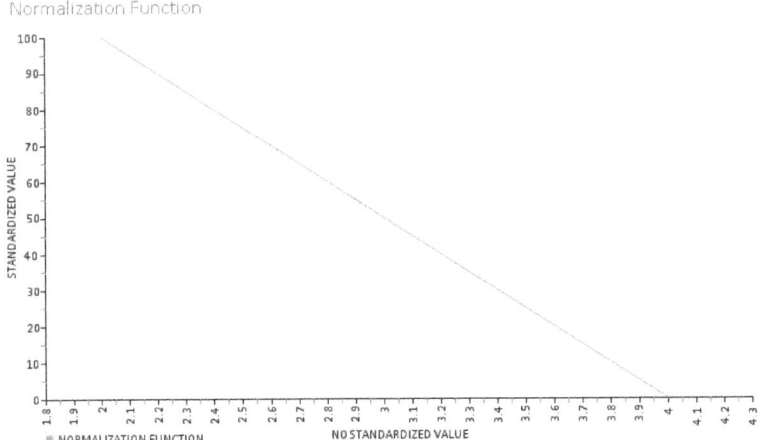

Variables

[AS-V5.1] Monthly expenditure per household on drinking water and wastewater services
Definition: Monthly expenditure per household on drinking water and wastewater services. Expenditure is calculated by assuming daily drinking water consumption of 200 liters, or monthly consumption of 6 cubic meters, and by applying the tariff for this volume of consumption/wastewater discharge in force at the end of the calendar year preceding the rating date. If it exists, the tariff in force for low-income households will be used when calculating expenditure. If low-income households receive a government subsidy paid directly to the user or discounted from the bill, this subsidy will be subtracted when calculating expenditure. If the "geographical area to be rated" includes several tariff zones, the tariff applied to the zone with the greatest number of users will be used.
Units: local currency
Reliability: Table 103

[AS-V5.2] Monthly expenditure per household on the drinking water service
Definition: Monthly expenditure per household on the drinking water service. Expenditure is calculated by assuming daily drinking water consumption of 200 liters, or monthly consumption of 6 cubic meters, and by applying the tariff for this volume of consumption in force at the end of the calendar year preceding the rating date. If it exists, the tariff in force for low-income households will be used when calculating expenditure. If low-income households receive a government subsidy paid directly to the user or discounted from the bill, this subsidy will be subtracted when calculating expenditure. If the "geographical area to be rated" includes several tariff zones, the tariff applied to the zone with the greatest number of users will be used.
Units: local currency
Reliability: Table 103

[AS-V6] Average monthly income per household in the poorest quintile.
Definition: Average monthly income per household in the poorest quintile.
Units: local currency
Reliability: Table 104

CG Corporate Governance

The quality of corporate governance of the organization or company (utility) responsible for delivering the services plays a decisive role in the directors' decision-making processes and affects overall utility performance. Consequently, it is one of the areas assessed by AquaRating. In this context, corporate governance is understood to be the relationship between the utility's directors and its owners, and between the directors and other stakeholders.

Autonomy, accountability and transparency are the key concepts and criteria assessed in this area, all of which are relevant to the quality of the utility's corporate governance, whether publicly or privately owned. Autonomy (associated with responsibility), accountability and transparency are conditions which, when fully met, create incentives for managers to run the utility effectively and according to established objectives and enable stakeholders to monitor management.

The sub-areas of evaluation are:

    CG1 Utility autonomy and responsibilities
    CG2 Decision-making processes and accountability
    CG3 Transparency and control

CG1 Utility autonomy and responsibilities

Utility autonomy, when associated with the responsibilities defined, transparency and accountability, contributes towards effective management. In a context in which service delivery objectives and associated responsibilities are clearly defined, management autonomy to decide about key aspects of management, such as staffing and remuneration, procurement, making payments, incurring debt, etc., will facilitate effective utility management and help it to achieve its objectives. This condition is assessed via good practices.

Practices

CG1.1   Utility autonomy and responsibilities

CG1.1 Utility autonomy and responsibilities

Type: Best Practices
Service: Drinking Water and/or Sanitation
Normalization: Weighted by practices
Glossary: Geographical service coverage area, Owner (utility), Legal person(ality), Formal autonomy (to acquire goods and services and make payments, to set remunerations and determine staffing; and to contract debt)., Business autonomy (for decisions related to remunerations and staffing, to acquisition of goods and services, and to debt), Board of directors
Definition: Includes:

Practices	Reliability	Weight
1	The utility is a "legal person" governed by either public or private law and is separate from the authority that awarded the service.	T. 105	1
2	Legal instruments exist that clearly establish the "geographic service coverage area" and the utility's rights and obligations regarding: service nature, service quality, service expansion or coverage goals, tariffs, investment plans and financing of investment.	T. 105	1
3	Articles of incorporation and bylaws or similar valid instruments exist that establish the structure, functions and responsibilities of the utility's decision-making bodies (e.g. meetings of "owners", "board of directors", etc.) and regulate these bodies' decision-making processes.	T. 105	1
4	The utility has "formal autonomy" to acquire goods and services, to make payments, to set remunerations and to determine staffing.	T. 105	1
5	The utility has "formal autonomy" to contract national debt.	T. 105	1
6	The utility has "business autonomy" for decisions related to remunerations and staffing and to procurement and indebtedness.	T. 105	2

CG2 Decision-making processes and accountability

Accountability to representatives of utility owners and other stakeholders is an important condition that contributes towards ensuring that those who have the duty and power to do so manage third-party resources responsibly and in accordance with the corporate governance rules established for the organization. A prior condition for accountability is clear definition of duties and decision-making processes. These aspects are assessed via the good practices described below, which are complemented by assessment of criterion CG3 as regards providing transparent information for the purpose of ensuring accountability to stakeholders in the widest sense, such as users, citizens, tax authorities, etc.

Practices

CG2.1   Corporate governance
CG2.2   Selection of "board" members and chief executive officer
CG2.3   "Board of directors'" powers and responsibilities

CG2.1 Corporate governance

Type: Best Practices
Service: Drinking Water and/or Sanitation
Normalization: Weighted by practices
Glossary: Owner (utility), Board of directors, Body that represents the owners, Timely reception of notices (board members), Acceptable timeframe for circulation of minutes issued by the board, Timely reception of notices (members of body that represents the owners), Acceptable timeframe for circulation of minutes issued by the body that represents the owners
Definition: Includes:

Practices	Reliability	Weight
1 A "body exists that represents the owners", this body being different to the "board of directors" in terms of composition and assigned powers.	T. 105	2
2 The main corporate "bodies" (the body that represents the owners, the "board of directors", and other bodies mentioned in the utility's articles of incorporation and bylaws) meet in accordance with the applicable regulations/articles of incorporation and bylaws and their decisions are recorded in formal minutes which accredit such meetings in time and form.	T. 4	2
3 All members of the "board of directors" and of its committees, in the case that such committees exist, receive "timely" notice of and agendas for all meetings, allowing them to attend and prepare properly.	T. 4	1
4 Documents which transcribe decisions that have been made (records, minutes) in meetings of the "board of directors" and its committees, in the case that such committees exist, are circulated among the members of the board of directors and its committees within an "acceptable timeframe" and are validated according to the rule established to this effect.	T. 4	1
All members of the "body that represents the owners" receive "timely" notice of and agendas for all meetings, allowing them to attend and prepare properly. (This practice is considered complied with at maximum reliability level if such body does not exist, as acknowledged by selecting		

5   non-compliance with practice CG2.1.1).   T. 4   1

6   Documents which transcribe decisions that have been made (records, minutes) in meetings of the "body that represents the owners", are circulated among its members and among all the members of the "board of directors" and its committees, in the case that such committees exist, within an "acceptable timeframe". (This practice is considered complied with at maximum reliability level if such body does not exist, as acknowledged by selecting non-compliance with practice CG2.1.1).   T. 4   1

7   The utility has a written and duly approved set of corporate governance policies that at the very least address "owners'" rights and relationship with the utility, the role of the "board of directors", the decision-making processes and communication of the decisions made.   T. 2   1

8   A unit exists that, among its functions, supervises development of and compliance with corporate governance policies.   T. 105   1

9   The utility has a written code of ethics approved by the "board of directors" that includes measures to prevent and detect corruption and is signed by all members of the board of directors and by all staff.   T. 2   1

10   A unit exists which, among its functions, ensures compliance with the code of ethics.   T. 105   1

11   A policy exists and is applied to promote public participation including, among other elements, meetings or consultations with citizens' representatives at least once a year.   T. 2   1

CG2.2 Selection of "board" members and chief executive officer

Type: Best Practices
Service: Drinking Water and/or Sanitation
Normalization: Weighted by practices
Glossary: Board of directors, Independent director
Definition: Includes:

	Practices	Reliability	Weight
1	Directors are chosen on the basis of a previously defined process that assesses their professional background and qualifications for the position.	T. 106	2
2	The chief executive officer (CEO) is chosen on the basis of a previously defined process that assesses his/her professional background and qualifications for the position.	T. 106	2
3	The positions of president of the "board" and chief executive officer (CEO) are held by two different persons.	T. 106	2
4	The members of the "board of directors" are elected or confirmed in their position at least every two years.	T. 107	1
5	When selecting members of the "board of directors" and the chief executive officer (CEO), either a firm of consultants specializing in recruiting executives is used or a board committee specializing in executive appointments is set up.	T. 106	1
6	The articles of incorporation and bylaws consider appointment of "independent directors" and at least 20% of "board" members are independent.	T. 106	1
7	Dismissal of a director is only possible on grounds explicitly stated in the articles of incorporation and bylaws or in equivalent documents.	T. 106	1

CG2.3 "Board of directors'" powers and responsibilities

Type: Best Practices
Service: Drinking Water and/or Sanitation
Normalization: Weighted by practices
Glossary: Owner (utility), Board of directors, Independent director
Definition: Includes:

	Practices	Reliability	Weight
1	The "board of directors" is the sole body responsible for selecting the chief executive officer (CEO) and determining his/her remuneration.	T. 106	1
2	The "board of directors" approves and monitors implementation of the utility's strategic plans and policies.	T. 2	1
3	The "board of directors" assesses the outcomes of management of the utility.	T. 2	1
4	The "board of directors" has established committees for specific issues, among which there is at least one of the following: auditing, remuneration, and appointment of directors and executives.	T. 106	1
5	The committee or committees mentioned in the former practice comprise a majority of "independent directors".	T. 106	1
6	The "board of directors" supervises directly or through the audit committee the utility's internal and external audits.	T. 2	1
7	The "board of directors" is held accountable once a year to the meeting of "owners" or to another body that represents public interests (in the case of public utilities).	T. 2	2
8	The "board of directors" self-assesses its performance	T. 2	1

CG3 Transparency and control

Transparency is necessary to make accountability effective and to hold the utility responsible for the outcomes of its management, neither of which are possible without ordered, timely and publicly available information (transparent information). Accounting and financial information play an important role, as transparency of these is essential to assessing utility management and exercising public control. This condition is assessed via good practices.

Practices

CG3.1   Disclosure of information about service delivery
CG3.2   Disclosure of institutional and financial information
CG3.3   Auditing and control processes

CG3.1 Disclosure of information about service delivery

Type: Best Practices
Service: Drinking Water and/or Sanitation
Normalization: Weighted by practices
Glossary: Applicable regulations, Complaint
Definition: As a minimum, the utility discloses or makes public the following:

Practices	Reliability	Weight
1 Tariffs and prices in force.	T. 108	1
2 User rights.	T. 108	1
3 Utility obligations regarding service quality (drinking water and/or wastewater).	T. 108	1
4 Compliance with applicable drinking water "standards".	T. 108	1
5 Information related to service delivery, such as service interruptions, works on public thoroughfares, etc.	T. 108	1
6 Number of "complaints" and time taken to resolve them.	T. 108	1

CG3.2 Disclosure of institutional and financial information

Type: Best Practices
Service: Drinking Water and/or Sanitation
Normalization: Weighted by practices
Glossary: Owner (utility), Board of directors, Body that represents the owners, Works (in relation to investment plan projects)
Definition: As a minimum, the utility discloses or makes public the following:

	Practices	Reliability	Weight
1	Its corporate governance policy, its degree of compliance with effectively implemented corporate governance policies, and the extent to which such practices conform to the country's voluntary code of good corporate governance.	T. 108	1
2	The utility's code of ethics, duly approved.	T. 108	1
3	Its financial statements and annual reports, within the regulated timeframe.	T. 108	3
4	Audit reports of external bodies.	T. 108	1
5	Information about who are the "utility's owners" and who are the members of (i) the "body that represents the owners" and (ii) the "board of directors".	T. 108	1
6	Remuneration, fees or allowances of each member of the utility's "board of directors".	T. 108	1
7	Investment plan and invitations to tender for programmed "works".	T. 108	1
8	Summary of all contracts that exceed the amount established to this effect in the corresponding regulations (i.e. the utility complies with its legal obligation to make public all contracts awarded).	T. 108	1

CG3.3 Auditing and control processes

Type: Best Practices
Service: Drinking Water and/or Sanitation
Normalization: Weighted by practices
Glossary: Board of directors, Body that represents the owners
Definition: Includes:

Practices	Reliability	Weight
1 The external auditors are chosen by the "body that represents the owners" on recommendation of the "board of directors", or the external auditors are chosen by the official entity responsible for external financial control of the utility.	T. 107	1
2 The "board of directors" or its audit committee examines and approves the external auditors' reports and those of the internal audit unit.	T. 2	1
3 The "board of directors" or its audit committee supervises implementation of recommendations made by external auditors and by the internal audit unit.	T. 2	1
4 An internal audit unit exists that reports to the board of directors/"board of directors'" audit committee.	T. 2	1

ES Environmental Sustainability

This rating area assesses the environmental sustainability of the way the utility manages the "systems" for it is responsible. It examines the degree of implementation of environmental considerations in system management and the environmental impacts that it produces. To quantify these impacts, the parameters most frequently used to characterize aquatic environments and their related ecosystems are employed.

This area only considers interaction between the environment and the possible impacts on it produced by activities linked to the rated drinking water and wastewater utilities. Assessment is limited to an environmental point of view and, consequently, to evaluation of the effects produced in the environment within current time horizons and to possible future repercussions.

Assessment is structured into two sub-areas: the first centers on the environmental impact deriving from the conditions of return to the aquatic environment of used water, or of water drained from the geographical area to be rated; the second considers all quantifiable interactions with the environment during planning and management of the services assessed, as well as the by-products and impacts generated.

The sub-areas of evaluation are:

ES1 Wastewater treatment and management
ES2 Environmental management

ES1 Wastewater treatment and management

This sub-area focuses on activities related to wastewater treatment processes because of their significant direct impact on the aquatic environment in water bodies receiving urban wastewater and rainwater collected by urban sewerage and drainage networks.

Practices

ES1.1 Assurance of operation and control of wastewater treatment services

Indicators

ES1.2 Availability of operational wastewater treatment infrastructure
ES1.3 Degree of compliance with discharge regulations

ES1.1 Assurance of operation and control of wastewater treatment services

Type: Best Practices
Service: Drinking Water and/or Sanitation
Normalization: Weighted by practices
Glossary: Real time, Preventive maintenance, Corrective maintenance, Preventive maintenance protocol, Corrective maintenance protocol, Wastewater treatment plan
Definition: Includes:

	Practices	Reliability	Weight
1	An up-to-date "wastewater treatment plan" is in force, compliance with which is controlled.	T. 2	2
2	The treatment capacity of operational wastewater treatment plants is greater than or equal to the maximum loads and flows they receive.	T. 8	3
3	Stormwater tanks exist to buffer peak pollutant flows from the urban drainage system and these tanks are equipped with a system to manage treatment of the volumes of stormwater stored in the wastewater treatment plants.	T. 3	1
4	"Protocols" and records exist for "preventive maintenance" performed on wastewater treatment plants.	T. 2	3
5	"Protocols" and records exist for "corrective maintenance" performed on wastewater treatment plants.	T. 2	3
6	Wastewater treatment plants serving more than 5,000 inhabitants are equipped with automated operation systems.	T. 1	2
7	Protocols exist for effluent self-assessment, performed daily or more frequently than required by applicable regulations.	T. 2	3
8	"Real-time" inflow and outflow gauging equipment exists.	T. 1	1
9	Equipment exists to measure physical and chemical parameters at inlet and outlet and in intermediate processes.	T. 1	2

A record of measurements and of operating parameter

| 10 | controls (flows and parameters stipulated by standards) is kept. | T. 2 | 1 |
| 11 | Water quality analysis laboratories (to analyze fundamental parameters) exist in at least 25% of wastewater treatment plants. | T. 8 | 1 |

ES1.2 Availability of operational wastewater treatment infrastructure

This assessment element considers the degree of wastewater treatment coverage for wastewater generated in the "geographical area to be rated". Coverage is measured as the percentage of "population equivalent" whose wastewater is treated in an operational wastewater treatment plant and receives at least secondary treatment. In cases in which wastewater is discharged via marine outfalls, this requirement will be considered met if the outfall complies with applicable national regulations.

Definition: Percentage of properties served (in "population equivalent") in the "geographical area to be rated" for wastewater collection whose discharges are connected to an operational treatment plant at the end of the calendar year preceding the rating date.
Type: Indicator
Service: Sanitation
Glossary: Population equivalent, Geographical area to be rated
Formula: ([SA-V9]/[SA-V14])*100   Unit: %
Normalization Function:

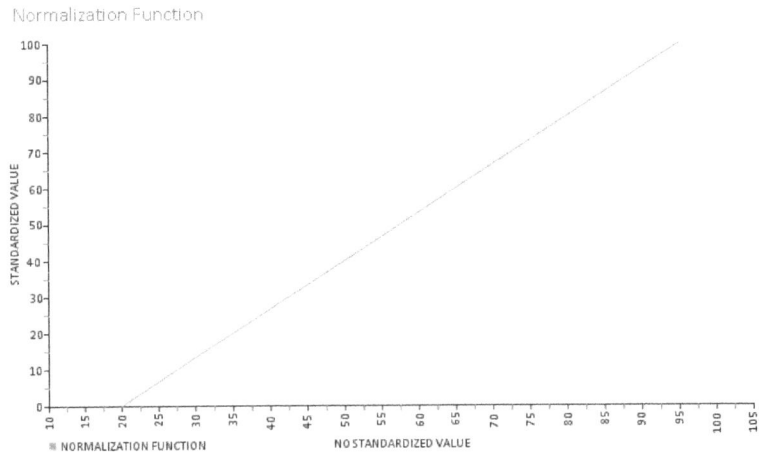

Variables

[SA-V14] Population "equivalent" that discharges wastewater within the "geographical area to be rated" for wastewater collection

Definition: Population "equivalent" that discharges wastewater within the "geographical area to be rated" for wastewater collection at the end of the calendar year preceding the rating date.
Units: population equivalent
Reliability: Table 117

[SA-V9] Properties whose discharges are connected to an operative wastewater treatment plant

Definition: Properties (in "population equivalent") whose discharges are connected to an operative wastewater treatment plant at the end of the calendar year preceding the rating date.
Units: no.
Reliability: Table 116

ES1.3 Degree of compliance with discharge regulations

This assessment element considers wastewater treatment "system" quality as a function of compliance with "applicable regulations" regarding the characteristics of effluent discharged by wastewater treatment plants in the system rated.

Definition: Number of samples taken that comply with "applicable regulations" as a proportion of all control samples taken. If the number of samples taken is less than the number established in the applicable regulations, this required number will be used as the indicator's denominator.
Type: Indicator
Service: Sanitation
Glossary: Applicable regulations, System
Formula: ([SA-V10]/[SA-V11])*100   Unit: %
Normalization Function:

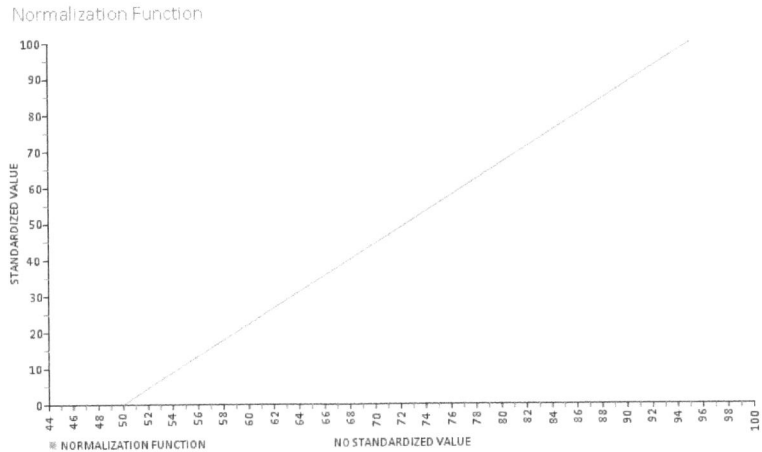

Variables

[SA-V10] Number of standard-compliant samples taken
Definition: Number of standard-compliant samples taken (compliant samples taken in the full calendar year preceding the rating). Both samples taken for self-assessment as well as those taken by entities responsible for monitoring and control will be considered.
Units: no.
Reliability: Table 118

[SA-V11] Total number of control samples taken and analyzed
Definition: Total number of control samples taken and analyzed (in the calendar year preceding the rating) or number of samples established in "applicable regulations" (the higher of the two values).
Units: no.
Reliability: Table 118

ES2 Environmental management

In this case, assessment is structured into three groups of practices that include planning, operation and management of "systems" and services. It also includes six quantitative measures of environmental impact: alterations to water resource flows, energy usage, greenhouse gas emissions, sludge generation in treatment processes, degree of water resource usage, and compliance with environmental regulations.

Practices

ES2.1   Environmental management framework
ES2.2   Environmental implications in planning
ES2.3   Environmental operation and promotion

Indicators

ES2.4   Water withdrawal in relation to the renewable resource
ES2.5   Energy consumption balance
ES2.6   Greenhouse gas emissions linked to drinking water and/or wastewater management
ES2.7   Environmental management of sludge produced by treatment processes
ES2.8   Water resource use
ES2.9   Compliance with environmental regulations

ES2.1 Environmental management framework

Type: Best Practices
Service: Drinking Water and/or Sanitation
Normalization: Weighted by practices
Glossary: System
Definition: Includes:

Practices	Reliability	Weight
1	The utility has a department or unit that deals exclusively with environmental issues.	T. 6	1
2	Commitments, protocols or internal procedures with a clear environmental focus exist.	T. 6	2
3	Drinking water supply and/or wastewater collection and/or treatment facilities (according to the services to be rated) hold certification to a recognized environmental standard, such as ISO 14001. Certification extends to facilities that represent at least 80% of installed capacity in the geographical area to be rated or 80% of the length of linear systems (pipes or sewers). Certification will always extend to the following: wastewater "systems" and wastewater treatments plants handling >300 m3/hour, facilities linked to the drinking water supply "system" handling >100 l/s.	T. 6	3

ES2.2 Environmental implications in planning

Type: Best Practices
Service: Drinking Water and/or Sanitation
Normalization: Weighted by practices
Glossary: System, Recycled/Recycling, Water bodies
Definition: Includes:

Practices	Reliability	Weight
1 Strategic planning establishes environmental goals and mechanisms for following up and monitoring deviation.	T. 2	3
2 Explicit plans or policies exist that address climate change mitigation or adaptation to climate change.	T. 2	2
3 Strategic or managerial objectives include compliance with goals for wastewater discharge quality that are more stringent than those established by law.	T. 2	2
4 Environmental and social costs are considered in assessment of all alternatives (initiatives, projects and works) planned or in planned operation.	T. 2	2
5 Planning considers wastewater reuse systems and the decision is substantiated by complete comparative analyses.	T. 2	2
6 Individual or collective "recycling" systems and use of non-conventional water sources are promoted or financed, provided their efficiency is substantiated.	T. 2	2
7 Mechanisms to facilitate public participation (civil society, environmental groups) in decisions that have environmental and social implications are considered.	T. 2	1
8 In the area influenced by management of the drinking water and/or wastewater "system" to be rated, "bodies of water" and ecosystems linked to them are classified on the basis of ecosystem quality and value in terms of biodiversity, singular species, area of high environmental value, etc.	T. 2	3

ES2.3 Environmental operation and promotion

Type: Best Practices
Service: Drinking Water and/or Sanitation
Normalization: Weighted by practices
Glossary:
Definition: Includes:

Practices	Reliability	Weight
1	Environmental and social impacts and hazards are assessed for all works and projects subject to such assessment under applicable legislation.	T. 2	2
2	A set of indicators to monitor and assess the environmental sustainability of system management is systematically applied.	T. 2	2
3	Environmental responsibility reports or equivalent documents are published systematically and include all environmental parameters habitually employed internationally, including, at the very least, those produced by the GRI (Global Reporting Initiative).	T. 2	2
4	As a minimum, environmental training programs exist for staff.	T. 2	1
5	Energy efficiency improvement programs exist.	T. 2	2
6	Water use efficiency and water demand management programs exist.	T. 2	2
7	An environmentally responsible culture is promoted.	T. 2	1
8	Research projects with principally environmental aims are financed, promoted or implemented.	T. 2	1

ES2.4 Water withdrawal in relation to the renewable resource

This element intends to reflect how water intake and distribution for urban supply modify, in each assessed system, the natural inflows and outflows in the watershed and in the water body used. The term renewable applies mainly to those cases in which groundwater is used, in which case modification refers to the relationship between the volumes withdrawn from the aquifer and natural aquifer recharge.

In many cases, water withdrawals from an abundant natural environment will not have an appreciable impact on ecosystems or on the status of water bodies. However, in other cases, this impact might be substantial, hence the interest in including an indicator that assesses the degree of modification of natural flows (even though this activity constitutes the foundations on which supply of water to urban areas is based).

This indicator measures the relative influence of withdrawal for water supply, irrespective of whether other agents affect natural flow balance. Possible overlap with other activities is not taken into consideration.

Definition: Percentage representing the volume of water withdrawn annually from the natural aquatic environment, both directly and imported from other systems, for incorporation in the water supply "system" in the "geographical area to be rated" (the average for the last five full years preceding rating is considered) as a proportion of the average annual natural inflow.
Type: Indicator
Service: Drinking Water
Glossary: System, Geographical area to be rated
Formula: ([SA-V1]/[SA-V2])*100   Unit: %
Normalization Function:

AquaRating

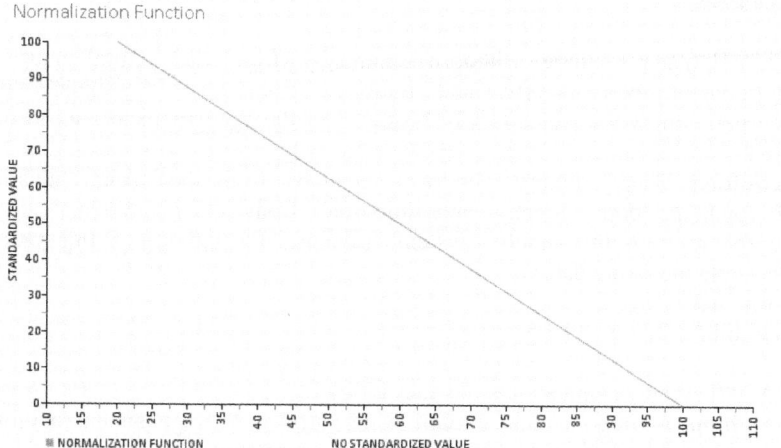

Normalization Function

STANDARDIZED VALUE

NORMALIZATION FUNCTION   NO STANDARDIZED VALUE

Variables

[SA-V1] Water volume withdrawn annually from the natural aquatic environment, both directly and imported from other "systems", for incorporation in the water supply "system" in the "geographical area to be rated"
Definition: Water volume withdrawn annually from the natural aquatic environment, both directly and imported from other "systems", for incorporation in the supply "system" (average for the last 5 full years preceding the rating date).
Units: m3
Reliability: Table 109

[SA-V2] Average annual natural water inflow
Definition: Average annual natural water inflow Average annual natural inflows and groundwater recharges will be considered. The average of the time series available for a period of at least 5 years preceding the rating will be taken.
Units: m3
Reliability: Table 110

ES2.5 Energy consumption balance

This indicator considers, from an overall perspective, the impact that energy consumption has on the environment (regardless of its type or degree of usage efficiency). As energy consumption depends largely on context and degree of compliance with service and environmental standards, an assessment element that considers the balance between energy produced and energy used is applied, assuming that energy may be produced as part of water and wastewater service delivery. The degree of energy usage efficiency is included in the Operating Efficiency rating area.

Definition: Percentage representing the energy consumed in all drinking water and wastewater service processes as a proportion of the energy generated in all facilities linked to the "system". The average annual value for the 3 full calendar years preceding the rating is used.
Type: Indicator
Service: Drinking Water and/or Sanitation
Glossary: System
Formula: ([SA-V3]/[SA-V4])*100   Unit: %
Normalization Function:

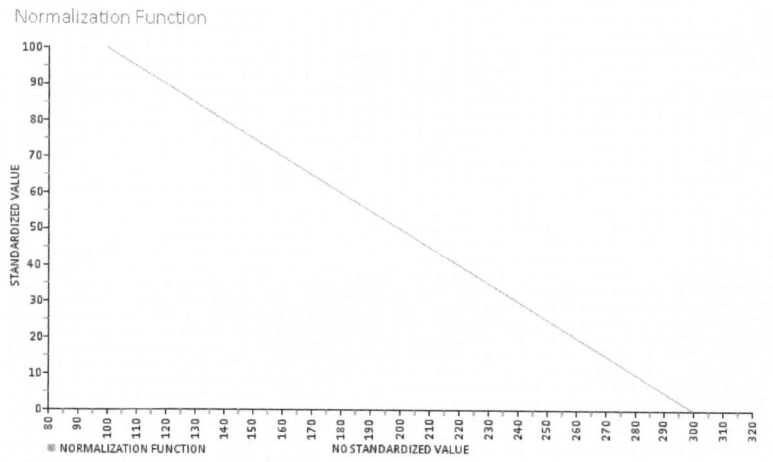

Variables

[SA-V3] Energy consumed by all drinking water and wastewater processes
Definition: Energy consumed by all drinking water and wastewater processes (average annual value for the last 3 full calendar years preceding the rating).
Units: Kwh
Reliability: Table 111

[SA-V4] Energy generated in facilities linked to the "system"
Definition: Energy generated in facilities linked to the "system" (average annual value for the last 3 full years).
Units: Kwh
Reliability: Table 112

ES2.6 Greenhouse gas emissions linked to drinking water and/or wastewater management

This indicator considers the environmental impact linked to greenhouse gas emissions. It is a variable that depends on the types of processes employed and rating scope. Nevertheless, this assessment element considers the potential environmental impact linked to the delivery of the rated service. In order to obtain more homogeneous values for this indicator, it is expressed as a proportion of the number of inhabitants served.

Definition: Annual tons of CO2 equivalent emitted per 1,000 inhabitants served
Type: Indicator
Service: Drinking Water and/or Sanitation
Glossary: Geographical area to be rated
Formula: ([SA-V5]/[SA-V15])*1000   Unit: Tons/1000 population
Normalization Function:

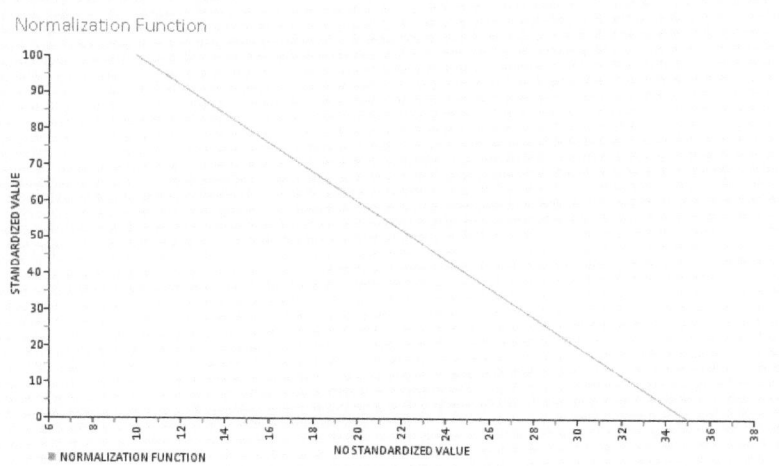

Variables

[SA-V15] Population served in the "geographical area to be rated"
Definition: Number of inhabitants receiving any of the services being rated in the "geographical area to be rated" at the end of the calendar year preceding the rating date.
Units: inhabitants
Reliability: Table 100

[SA-V5] Annual emission of CO_2 equivalent
Definition: Annual emission of CO_2 equivalent in the calendar year preceding the rating date.
Units: tons
Reliability: Table 113

ES2.7 Environmental management of sludge produced by treatment processes

This indicator assesses the destination and potential impact of sludge generated in both wastewater treatment processes and drinking water treatment processes.

The amount of sludge generated depends on the type of process and the type of wastewater treated. Therefore, this assessment element considers the proportion of all sludge generated assigned an "environmentally responsible destination".

Definition: Percentage of sludge used for energy production or assigned an "environmentally responsible destination".
Type: Indicator
Service: Drinking Water and/or Sanitation
Glossary: System, Environmentally responsible destination
Formula: ([SA-V6]/[SA-V7])*100   Unit: %
Normalization Function:

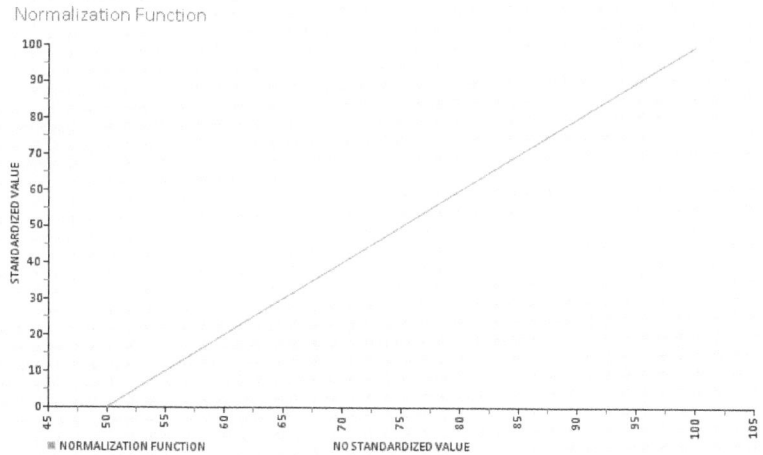

Variables

[SA-V6] Amount of sludge used for energy production or assigned an "environmentally responsible destination"
Definition: Amount of sludge used for energy production or assigned an environmentally responsible destination (full calendar year preceding the rating).
Units: tons
Reliability: Table 114

[SA-V7] Sludge produced by "system" processes
Definition: Sludge produced by "system" processes (full calendar year preceding the rating).
Units: tons
Reliability: Table 115

ES2.8 Water resource use

This assessment element considers the degree of environmental modification produced by unitary consumption of water resources. It is an indicator that depends greatly on contextual factors such as local climate, water use culture and commercial and industrial activity type and intensity in the rating area. Nevertheless, it is an indicator that enables evaluation of this set of factors in relation to water volume used as a proportion of the population served in the geographical area rated. It would not be defensible as an efficiency indicator because of the influence of all the contextual elements. However, as an environmental assessment element, it enables evaluation of the degree of impact produced by activities in the zone and by usage and consumption habits.

Definition: Water volume per inhabitant and day withdrawn from the natural environment for water supply (average of the 3 years preceding the rating).
Type: Indicator
Service: Drinking Water
Glossary: Geographical area to be rated
Formula: ([SA-V8]/[CS1-V2])   Unit: l/hab day
Normalization Function:

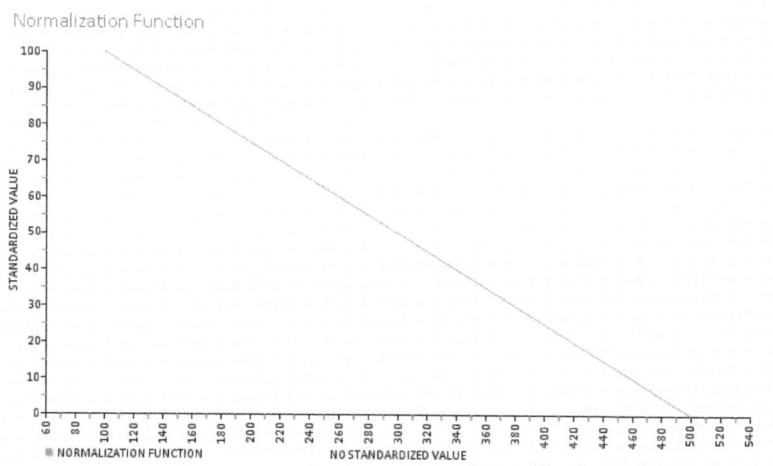

Variables

[CS1-V2] Population with a household connection in the "geographical area to be rated" for drinking water supply.
Definition: Population with a household connection in the "geographical area to be rated" for drinking water supply (at the end of the calendar year preceding the rating date).
Units: inhabitants
Reliability: Table 100

[SA-V8] Water volume withdrawn from the natural environment for water supply (daily)
Definition: Water volume withdrawn from the natural environment for water supply (daily)
Units: liters
Reliability: Table 109

ES2.9 Compliance with environmental regulations

This assessment element reflects the degree of compliance with the applicable environmental regulations in the area rated. It has the advantage of using an accessible and easily measured and controlled variable. However, it depends heavily on the extent of institutional monitoring and control performed in each case and circumstance. Regulations and monitoring will be different in each "system".

Definition: Percentage of "monitoring points" established by environmental regulations reported, investigated or fined for breaching "applicable regulations" in the calendar year preceding the rating date. Cases of non-compliance detected in ISO 14001 audits will also be considered breaches.
Type: Indicator
Service: Drinking Water and/or Sanitation
Glossary: Applicable regulations, System, Monitoring points
Formula: ([SA-V12]/[SA-V13])*100   Unit: %
Normalization Function:

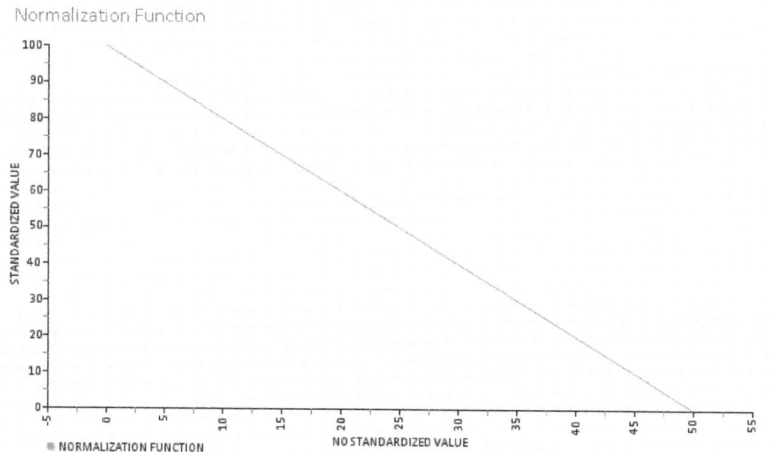

Variables

[SA-V12] Number of "monitoring points" reported, investigated or fined for breaching "applicable regulations"
Definition: Number of "monitoring points" reported, investigated or fined for breaching "applicable regulations" (in the calendar year preceding the rating date).
Units: no.
Reliability: Table 119

[SA-V13] Total number of "monitoring points" as per applicable environmental "regulations"
Definition: Total number of "monitoring points" as per applicable environmental "regulations" at the end of the calendar year preceding the rating date.
Units: no.
Reliability: Table 119

Appendix A - Reliability Tables

Table 1

Reliability Levels	Factor
1 Existence of equipment cannot be verified.	0
2 Equipment exists physically, it can be verified that it has the indicated characteristics and that the rated utility is authorized to use it (property, invoice, delivery note, rent receipt or other evidence).	0.6
3 In addition to fulfilling the requirements for level 2, operation manuals and personnel trained to use and maintain the equipment exist.	0.7
4 In addition to fulfilling the requirements for levels 2 and 3, calibration of at least 60% of the equipment is certified (by an accredited laboratory in the case of equipment for measuring physical and chemical parameters) and accuracy is systematically verified.	0.8
5 In addition to the fulfillment of requirements for levels 2 and 3, for 100% of the equipment, calibration is certified (by an accredited laboratory in case of equipment for measuring physico-chemical parameters) and accuracy is systematically verified.	0.95

6	In addition to fulfilling the requirements for levels 2, 3 and 5, it is part of the instrumentation audited, at least internally, according to an accreditation standard.	1

Table 2

	Reliability Levels	Factor
1	The practice is not documented.	0
2	The practice is documented, but there is no evidence of its application either at the rating date or in the calendar year preceding that date.	0.5
3	The practice is documented and evidence exists of its application at the rating date or in the calendar year preceding that date.	0.7
4	The practice is documented and evidence exists of its application at the rating date as well as in the calendar year preceding that date or in the previous 2 calendar years.	1

Table 3

	Reliability Levels	Sum
1	Documents describing the system exist, as do handbooks for its use and maintenance.	0.25
2	Staff are available to use and maintain it.	0.25
3	It is permanently installed on all relevant workstations or can be accessed from them.	0.25
4	Records of its systematic use exist.	0.25

Table 4

Reliability Levels	Factor
1 There is no evidence of its application.	0
2 Evidence exists of its application at the rating date or in the calendar year preceding that date.	0.7
3 Evidence exists of its application at the rating date as well as in the calendar year preceding that date or in the 2 calendar years preceding the rating date.	1

Table 5

Reliability Levels	Factor
1 There is no evidence of its application.	0
2 Evidence exists of its application in the calendar year preceding the rating date.	0.7
3 Evidence exists of its application at the rating date.	0.8
4 Evidence exists of its application at the rating date as well as in the calendar year preceding that date.	0.9
5 Evidence exists of its application at the rating date as well as in the 2 calendar years preceding that date.	1

Table 6

Reliability Levels	Factor
1	The practice is not documented.	0
2	The practice is documented, but there is no evidence of its application either at the rating date or in the calendar year preceding that date.	0.5
3	A documented procedure exists and evidence exists of its application in the calendar year preceding the rating date.	0.7
4	A documented procedure exists and evidence exists of its application at the rating date.	0.8
5	A documented procedure exists and evidence exists of its application at the rating date as well as in the preceding calendar year.	0.9
6	A documented procedure exists and evidence exists of its application at the rating date as well as in the 2 preceding calendar years.	1

Table 7

Reliability Levels	Factor
1	No record exists. Based on estimates.	0
2	Paper records exist of infrastructure, type, nominal capacity and hydraulic connections on land registry and commercial maps. No clear correlation exists with census data.	0.7
3	Paper records exist of infrastructure, type, nominal capacity and hydraulic connections on land registry and commercial maps. Sufficient correlation exists between the geographical areas served and census data.	0.9
4	Computerized records of infrastructure and connection exist in an up-to-date GIS, as do computerized records of daily consumption and treatment capacities. Sufficient correlation exists between the geographical areas served and census data.	1

Table 8

Reliability Levels	Factor
1	Existence of the facilities cannot be verified.	0
2	Facilities exist physically, they may be visited, it can be verified that they have the indicated characteristics and that the rated utility is authorized to use them (property, lease, or concession agreement).	0.6
3	In addition to fulfilling the requirements for level 2, operation protocols and staff trained to use and maintain the facilities exist.	0.7
4	In addition to fulfilling the requirements for levels 2 and 3, proper operation of all the facilities has been verified according to an established protocol.	0.95
	In addition to fulfilling the requirements for levels 2, 3 and 4, they are part of the facilities audited, at least internally, according	

5 to an accreditation standard.   1

Table 9

Reliability Levels	Factor
1 No record exists.	0
2 Sampling and analysis are documented in unsigned records without quality control.	0.33
3 Sampling and analysis are documented in signed records subject to traceability criteria and quality control.	0.8
4 Measurements are documented in signed records subject to traceability criteria and quality control. A reliable system exists to link samples to population or to the corresponding properties.	1

Table 10

Reliability Levels	Factor
1 No record exists.	0
2 Paper records exist of samples, analyses and consumption zones.	0.5
3 Computerized records of infrastructure and connections exist in an up-to-date GIS, as do computerized records of samples and analyses, quality requirements and accreditation by the laboratories that perform the analyses.	1

Table 11

Reliability Levels	Factor
1 No currently applicable supporting documents exist.	0
2 Currently applicable supporting documents exist, but there is no evidence in writing of their application.	0.5
3 Currently applicable supporting documents exist and there is evidence in writing of their application at the rating date or in one of the 2 calendar years preceding that date.	1

Table 12

Reliability Levels	Factor
1 No documented records exist.	0
2 Paper records exist of "incidents", complaints, programmed service stoppages and service interruptions (including estimates of the number of properties affected).	0.5
3 GIS records exist of "incidents", complaints, programmed service stoppages and service interruptions (including estimates of the number of properties affected).	0.8
4 GIS records exist of "incidents", complaints, programmed service stoppages and service interruptions (including precise details of the number of properties affected).	1

Glossary: Incident

Table 13

	Reliability Levels	Factor
1	No documented records other than census data exist.	0
2	Paper records of users and properties served exist.	0.5
3	Computerized records of users exist, indicating type and property for all users.	0.8
4	GIS records of connections exist, indicating associated properties and linked to the distribution network and the "incident" management system.	1

Glossary: Incident

Table 14

	Reliability Levels	Factor
1	No documented records exist.	0
2	Paper records exist of applications, completed connections and service availability.	0.7
3	Computerized records exist of applications, completed connections and service availability.	0.9
4	Computerized records exist of applications, processing, completion and service availability notification.	1

Table 15

Reliability Levels	Factor
1 No record exists. Based on estimates.	0
2 Paper records exist of receipt and resolution processes.	0.5
3 Computerized (alphanumeric) records exist of receipt and resolution processes.	0.8
4 Computerized records with GIS references exist and are linked to the "incident" and works management system.	1

Glossary: Incident

Table 16

Reliability Levels	Factor
1 No record exists. Based on estimates.	0
2 Paper records exist of processes and work orders up to resolution and user notification.	0.5
3 Computerized records exist of work orders carried out.	0.8
4 Computerized records exist of work orders carried out and user notifications. These include GIS references and are linked to the user management system.	1

Table 18

Reliability Levels		Sum
1	The survey is representative of the user population.	0.4
2	The survey methodology is stable and replicable.	0.2
3	The survey is conducted by a group/organization with appropriate "technical expertise".	0.2
4	The survey is conducted by a third party.	0.2

Glossary: Technical expertise

Table 19

Reliability Levels		Sum
1	Survey of all users who have experienced a problem, or of a representative sample of them.	0.4
2	The survey methodology is stable and replicable.	0.2
3	The survey is conducted by a group/organization with appropriate "technical expertise".	0.2
4	The survey is conducted by a third party.	0.2

Glossary: Technical expertise

Table 20

Reliability Levels	Factor
1 No "complaint" record exists.	0
2 Paper records exist of all "complaints".	0.7
3 Computerized records exist of "complaints".	1

Glossary: Customer service complaint

Table 21

Reliability Levels	Factor
1 No record exists.	0
2 Paper records.	0.4
3 Computerized records not integrated with the accounting system.	0.8
4 Computerized records integrated with the accounting system.	1

Table 22

Reliability Levels	Factor
1 No record exists	0
2 Handwritten/paper records kept by telephone operators.	0.6
3 Central telephone switchboard (PBX) with call waiting management.	1

Table 23

Reliability Levels	Factor
1 No record exists	0
2 "Real-time" electronic monitoring of waiting time during all office hours, without record segregation by consultation type.	0.67
3 "Real-time" electronic monitoring of waiting time during all office hours, with record segregation by consultation type.	1

Glossary: Real time

Table 24

Reliability Levels	Factor
1 No record exists	0
2 Paper records.	0.8
3 Computerized records.	0.95
4 Computerized records with evidence of user notification of resolution.	1

Table 25

Reliability Levels	Factor
1 No documentation exists.	0
2 Existing documentation was approved more than 5 years ago.	0.5
3 Existing documentation was approved less than 5 years ago.	1

AquaRating

Table 26

Reliability Levels	Factor
1 Documented drawings do not exist.	0
2 Existing documented drawings were approved more than 5 years ago.	0.3
3 Existing documented drawings were approved less than 5 years ago, but only partially fulfil the characteristics established.	0.7
4 Existing documented drawings were approved less than 5 years ago and fully fulfil the characteristics established.	1

Table 27

Reliability Levels	Factor
1 Cannot be corroborated.	0
2 Written evidence exists of its approval by the competent authority.	1

Table 29

Reliability Levels		Factor
1	No records exist of the investment plan, projects, contracts or their implementation.	0
2	Only overall records exist of projects and the investment plan.	0.3
3	Information about projects and investment costs is partial, or databases are fragmented, and there is no supporting documentation.	0.5
4	An integrated digital records system exists containing all relevant information about the plan's projects and their costs, but supporting documentation is partial or information on expenditure is not "consistent" with accounting records.	0.7
5	An integrated digital records system exists containing all relevant information about the investment plan, projects and works. It includes supporting documentation (contracts, physical and financial implementation, etc.) and information on expenditure is "consistent" with accounting records.	1

Glossary: Consistency of accounting information not originating from the financial statements

Table 30

Reliability Levels		Factor
1	No supporting documents exist. Only a list of works exists, which does not constitute an investment plan.	0
2	Existing documentation about the investment plan "in force" was approved more than 5 years ago.	0.3
3	Existing documentation about the investment plan "in force" was approved less than 5 years ago and includes a partial description of practices indicated in PE1.1.	0.7
4	Existing documentation about the investment plan was approved less than 5 years ago and includes a full description of the practices indicated.	1

Glossary: In force (investment plan in force)

Table 31

Reliability Levels		Factor
1	No systematized data records exist of final costs or of costs at time of award for "works finished" in the calendar year preceding the rating date.	0
2	Records of relevant information are partial or in fragmented databases and there are no supporting documents.	0.5
3	An integrated digital records system exists containing all relevant information, but documentary support is partial or costs of "finished works" are not "consistent" with accounting records.	0.7
4	An integrated digital records system exists containing relevant information about the "finished works". It includes supporting documentation (contracts, physical and financial implementation, etc.) and information on the cost of finished works is "consistent" with accounting records.	1

Glossary: Finished works, Consistency of accounting information not originating from the financial statements

Table 32

	Reliability Levels	Factor
1	No systematized data records exist of actual implementation timeframes or of implementation timeframes at time of award for "works finished" in the calendar year preceding the rating date.	0
2	Records of relevant information are partial or in fragmented databases and there are no supporting documents.	0.5
3	An integrated digital records system exists containing all relevant information, but documentary support is partial.	0.7
4	An integrated digital records system exists containing relevant information about implementation timeframes for "works finished" in the calendar year preceding the rating date. It includes supporting documentation (contracts, physical and financial implementation, etc.).	1

Glossary: Finished works

Table 33

	Reliability Levels	Factor
1	It cannot be corroborated or it was approved more than 4 years ago.	0
2	It is available and was approved between 3 and 4 years ago.	0.5
3	It is available and less than 3 years old, but it is not up to date at the rating date.	0.7
4	It is available and less than 3 years old and it is up to date at the rating date.	1

Table 34

	Reliability Levels	Factor
1	Cannot be corroborated.	0
2	Management of fixed assets is considered in the strategic plan's guidelines and there is a function related to this area in one of the organization's units.	0.6
3	Management of fixed assets is considered in the strategic plan's guidelines and objectives and a unit in the organization is currently in charge of this area.	1

Table 35

Reliability Levels	Factor
1 No separate accounting records exist for investment in replacement of or for expenditure on "corrective"/"preventive" maintenance of fixed physical assets or the data come from incomplete or unaudited financial statements or from audited financial statements receiving either a disclaimer of opinion related to this indicator or an adverse opinion.	0
2 Separate accounting records exist for investment in replacement of or for expenditure on "corrective"/"preventive" maintenance of fixed physical assets and financial statements are audited by external auditors ("registered" or non-registered), but definition or identification of these expenditures does not meet the "criteria established in the International Accounting Standards (IAS 16)" or the auditors' report contains a qualified opinion related to this indicator.	0.3
3 Separate accounting records exist for investment in replacement of or for expenditure on "corrective"/"preventive" maintenance of fixed physical assets and definition or identification of these expenditures meet the "criteria established in the International Accounting Standards (IAS 16)", but it is not possible to verify their "consistency<-90. with the financial statements audited by external auditors ("registered" or non-registered).	0.5
4 Separate accounting records exist for investment in replacement of or for expenditure on "corrective"/"preventive" maintenance of fixed physical assets, definition or identification of these expenditures meet the "criteria established in the International Accounting Standards (IAS 16)" and their amount is "consistent" with the financial statements audited by non-"registered" external auditors who have issued a disclaimer of opinion not related to this indicator.	0.7
Separate accounting records exist for investment in replacement of or for expenditure on "corrective"/"preventive" maintenance of fixed physical assets, definition or identification of these expenditures meet the "criteria established in the International Accounting Standards (IAS 16)" and their amount is "consistent" with the financial statements audited by "registered" external	

| 5 | auditors who have issued a disclaimer of opinion not related to this indicator. | 0.8 |

| 6 | Separate accounting records exist for investment in replacement of or for expenditure on "corrective"/"preventive" maintenance of fixed physical assets, definition or identification of these expenditures meet the "criteria established in the International Accounting Standards (IAS 16)" and their amount is "consistent" with the financial statements audited by non-"registered" external auditors who have issued an unqualified opinion related to this indicator. | 0.9 |

| 7 | Separate accounting records exist for investment in replacement of or for expenditure on "corrective"/"preventive" maintenance of fixed physical assets, definition or identification of these expenditures meet the "criteria established in the International Accounting Standards (IAS 16)" and their amount is "consistent" with the financial statements audited by "registered external auditors" who have issued an unqualified opinion related to this indicator. | 1 |

Glossary: Preventive maintenance, Corrective maintenance, Registered external auditors, Criteria established in the International Accounting Standards (IAS 16), Consistency of accounting information not originating from the financial statements

Table 36

Reliability Levels	Factor
1	The financial statements are incomplete or unaudited, or are audited and include either a disclaimer of opinion related to the fixed physical assets or an adverse opinion.	0
2	Financial statements audited by external auditors ("registered" or non-registered) that include either a qualified opinion related to the fixed physical assets, or ancillary records for the fixed assets "consistent" with the financial statements.	0.3
3	Financial statements audited by external auditors ("registered" or non-registered) that include either a disclaimer of opinion not related to the fixed physical assets, or ancillary records for the fixed assets "consistent" with the financial statements. The criterion for determining the fixed physical assets' value is based on the acquisition cost model, one of the options established in the International Accounting Standards.	0.5
4	Financial statements audited by external auditors ("registered" or non-registered) that include either a disclaimer of opinion not related to the fixed physical assets, or ancillary records for the fixed assets "consistent" with the financial statements. The criterion for determining the fixed physical assets' value is based on the revaluation model, which is the other option established in the International Accounting Standards.	0.6
5	Financial statements audited by non-"registered" external auditors that include an unqualified opinion related to the fixed physical assets, or ancillary records for the fixed assets "consistent" with the financial statements. The criterion for determining the fixed physical assets' value is based on the acquisition cost model, one of the options established in the International Accounting Standards.	0.7

Financial statements audited by "registered" external auditors that include an unqualified opinion related to the fixed physical assets, or ancillary records for the fixed assets "consistent" with the financial statements. The criterion for determining the fixed physical assets' value is based on the acquisition cost model, one of the options established in the International Accounting

6	Standards.	0.8
7	Financial statements audited by non-"registered" external auditors that include an unqualified opinion related to the fixed physical assets, or ancillary records for the fixed assets "consistent" with the financial statements. The criterion for determining the fixed physical assets' value is based on the revaluation model, which is the other option established in the International Accounting Standards.	0.9
8	Financial statements audited by "registered" external auditors that include an unqualified opinion related to the fixed physical assets, or ancillary records for the fixed assets "consistent" with the financial statements. The criterion for determining the fixed physical assets' value is based on the revaluation model, which is the other option established in the International Accounting Standards.	1

Glossary: Registered external auditors, Consistency of accounting information not originating from the financial statements

Table 37

Reliability Levels	Factor
1	The practice is not documented or it was reviewed and/or updated more than 3 years ago.	0
2	The practice is documented and there is evidence that it was reviewed and/or updated in the third calendar year preceding the rating date.	0.5
3	The practice is documented and there is evidence that it was reviewed and/or updated in the second calendar year preceding the rating date.	0.7
4	The practice is documented and there is evidence that it was reviewed and/or updated at the rating date or in the calendar year preceding the rating date.	0.9

The practice is documented and there is evidence that it was

5	reviewed and/or updated at the rating date, in the calendar year preceding the rating date or in the 2 years preceding the rating date.	1

Table 38

	Reliability Levels	Factor
1	The practice is not documented or it was reviewed and/or updated more than 3 years ago.	0
2	The practice is documented and there is evidence that it was reviewed and/or updated in the third calendar year preceding the rating date.	0.5
3	The practice is documented and there is evidence that it was reviewed and/or updated and implemented or incorporated into the investment plan, if applicable, in the second calendar year preceding the rating date.	0.7
4	The practice is documented and there is evidence that it was reviewed and/or updated and implemented or incorporated into the investment plan, if applicable, at the rating date or in the calendar year preceding the rating date.	0.9
5	The practice is documented and there is evidence that it was reviewed and/or updated and implemented or incorporated into the investment plan, if applicable, at the rating date, in the calendar year preceding the rating date or in the 2 calendar years preceding the rating date.	1

Table 39

Reliability Levels	Factor
1	The data is an estimate without any support in the utility's accounting system.	0
2	The data is estimated on the basis of ex-post classification of certain generic expenditures in the utility's accounting system, or on other substantiating documents.	0.6
3	The data come from the utility's accounting system, which segregates R&D costs, but the financial statements are not audited by external auditors.	0.7
4	The data come from the utility's accounting system, which segregates R&D costs (cost center, cost type), but their "consistency" with the information in the financial statements audited by external auditors ("registered" or non-registered) cannot be verified.	0.8
5	The data come from the utility's accounting system, which segregates R&D costs, and is "consistent" with the information in the financial statements audited by non-"registered" external auditors.	0.9
6	The data come from the utility's accounting system, which segregates R&D costs, and is "consistent" with the information in the financial statements audited by "registered" external auditors.	1

Glossary: Registered external auditors, Consistency of accounting information not originating from the financial statements

Table 40

Reliability Levels	Factor
1 No metering or micro-metering records exist.	0
2 Meter readings are recorded at least once a year.	0.33
3 Meter readings are recorded at least once a quarter. Real readings (without need for estimate) represent more than 90% of total readings.	0.9
4 Meter readings are recorded at least once every two months. Real readings (without need for estimate) represent more than 90% of total readings. Systematic practices exist to verify measurement reliability.	1

Table 41

Reliability Levels	Factor
1 No gauging records exist of intakes or introductions into the system.	0
2 Gauging records are taken at all "entry points" into the "system" at least once a year.	0.25
3 Gauging records are taken at all "entry points" into the "system" at least once a month.	0.75
4 Gauging records are taken at all "entry points" into the "system" at least once a day by remote monitoring systems.	0.9
5 Gauging records are taken at all "entry points" into the "system" at least once a day by remote monitoring systems. Gauging equipment calibration practices exist.	1

Glossary: System, Entry point into the drinking water supply system

Table 42

Reliability Levels	Factor
1	Real losses are estimated without access to any of the following information: data on water input into the "system", metered individual consumption (or calculated from a representative statistical base), or criteria for calculating uncontrolled water components.	0
2	Estimates based on balances and uncontrolled water components are made for the entire "system" at least once a year.	0.5
3	Estimates based on balances and uncontrolled water components are made once a month and calculations are substantiated by documented criteria or empirical references.	0.9
4	Estimates are based on balances and uncontrolled water components and calculations are substantiated by documented criteria or empirical references and compare equal reading periods for volumes supplied and consumed at the sector level.	1

Glossary: System

Table 43

Reliability Levels	Factor
1	No documented records exist.	0
2	Managed pipes are mapped on paper.	0.3
3	Managed pipes are recorded in a GIS. Systematic information maintenance and updating protocols do not exist.	0.8
4	Managed pipes are recorded in a GIS. Systematic information maintenance and updating protocols exist.	1

Table 44

Reliability Levels	Factor
1 No documented records exist.	0
2 Managed pipes are mapped on paper.	0.33
3 Computerized records exist of managed pipes and connections. Systematic maintenance and updating protocols do not exist.	0.66
4 Managed pipes and connections are recorded in a GIS. Systematic maintenance and updating protocols exist for information linked to customer management.	1

Table 45

Reliability Levels	Factor
1 No operation records exist.	0
2 Records of all operations performed on the infrastructure exist, but there is no data about their duration or the estimated volume used.	0.5
3 Records of all operations performed on the infrastructure exist, including data about their duration or the estimated volume used.	0.9
4 Records of all operations performed on the infrastructure exist, including data about their duration and the estimated volumes used based on reference measurements, working pressures and drain dimensions.	1

Table 46

Reliability Levels		Factor
1	No records for reclaimed or reused volumes exist.	0
2	Records and measurements of volumes reclaimed in wastewater reclamation plants exist, but there are no measurements of consumption at end destinations.	0.8
3	Records and measurements of reclaimed volumes exist, including measurements of consumption at end destinations.	1

Table 47

Reliability Levels		Factor
1	No evidence exists.	0
2	Evidence exists at the rating date or in one of the 4 calendar years preceding that date.	1

Table 48

Reliability Levels	Factor
1 No energy consumption records exist.	0
2 Energy consumption records exist for all wastewater treatment plants as a whole based on overall "system" records.	0.33
3 Energy consumption records exist for each wastewater treatment plant.	0.9
4 Energy consumption records exist for each wastewater treatment plant and are included in public documents or reports.	1

Glossary: System

Table 49

Reliability Levels	Factor
1 No records of pollutant loads in inflows and outflows exist.	0
2 Records of outflows and inflows exist for part of the year for all wastewater treatment plants.	0.33
3 Monthly outflow and inflow pollutant load records exist for all wastewater treatment plants.	0.9
4 Weekly or more frequent outflow and inflow pollutant load records exist for all wastewater treatment plants.	1

Table 50

Reliability Levels	Factor
1 No documented records exist.	0
2 Paper records exist of ruptures, "incidents" and repairs.	0.33
3 Computerized records exist of ruptures, "incidents" and repairs.	0.66
4 Records of distribution infrastructure, ruptures, "incidents" and repairs exist in a GIS and are classified by type, origin and responsibility.	1

Glossary: Incident

Table 51

Reliability Levels	Factor
1 No documented records exist.	0
2 Paper records exist of ruptures, "incidents" and repairs.	0.33
3 Computerized records exist of ruptures, "incidents" and repairs.	0.66
4 Records of connections, distribution infrastructure, ruptures, "incidents" and repairs exist in a GIS and are classified by type, origin and responsibility.	1

Glossary: Incident

Table 52

Reliability Levels	Factor
1 No record exists.	0
2 Paper records of "incidents" exist.	0.33
3 Computerized records of "incidents" exist.	0.8
4 Records of "incidents" and warnings exist in a GIS accessed from an incident and warning management center.	1

Glossary: Incident

Table 53

Reliability Levels	Factor
1 It cannot be corroborated or the plan was approved more than 5 years ago.	0
2 It is documented, but the plan was approved between 4 and 5 years ago.	0.5
3 It is documented and the plan was approved less than 4 years ago.	1

Table 54

Reliability Levels		Factor
1	It cannot be corroborated or there is evidence that not more than 20% of staff at the levels specified participated.	0
2	Evidence exists that between 20% and 50% of staff at the levels specified participated.	0.5
3	Evidence exists that more than 50% of staff at the levels specified participated.	0.8
4	Evidence exists that more than 80% of staff at the levels specified participated.	1

Table 55

Reliability Levels		Factor
1	Cannot be corroborated.	0
2	Written evidence exists that the plan was reviewed in at least one of the 3 calendar years preceding the rating date, if the plan was approved more than 3 years ago.	0.7
3	Written evidence exists that the plan was reviewed in the calendar year preceding the rating date, if the plan was approved more than 2 years ago.	0.9
4	Written evidence exists that the plan was reviewed in the last 2 calendar years preceding the rating date, if the plan was approved more than 2 years ago, or in the last calendar year preceding the rating date if the plan is 2 years old, or that the plan was formulated the year preceding the rating date.	1

Table 56

Reliability Levels	Factor
1 No supporting documents exist.	0
2 Supporting documents exist.	1

Table 57

Reliability Levels	Factor
1 No supporting documents exist.	0
2 Supporting documents exist, but there is no evidence of their application.	0.5
3 Supporting documents exist and evidence exists of their application.	1

Table 58

Reliability Levels	Factor
1 Cannot be corroborated.	0
2 It is documented, but there is no evidence of regular application in the 6 months preceding the rating date.	0.5
3 It is documented and evidence exists of regular application in the 6 months preceding the rating date.	0.9
4 It is documented and evidence exists of regular application at the rating date and in the previous calendar year.	0.95

It is documented and evidence exists of regular application at the

5 rating date and in the 2 previous calendar years.   1

Table 59

Reliability Levels	Factor
1 Cannot be corroborated.	0
2 A list of indicators used for management control exists indicating their objective and computation formula, or there are documents that describe the "management control system's" indicators, but there is no evidence of their application.	0.5
3 Evidence of evaluation of the management indicators exists at the rating date, but there are no documents describing the "management control system".	0.7
4 Documents describing the "management control system's" indicators exist specifying the objective, substantiation, computation formula, intervening variables and information sources for each one, and evidence exists of their evaluation at the rating date.	0.9
5 Documents describing the "management control system's" indicators exist specifying the objective, substantiation, computation formula, intervening variables and information sources for each one, and evidence exists of their evaluation at the rating date and in the previous calendar year.	0.95
6 Documents describing the "management control system's" indicators exist specifying the objective, substantiation, computation formula, intervening variables and information sources for each one, and evidence exists of their evaluation at the rating date and in the 2 previous calendar years.	1

Glossary: Management control system

Table 60

Reliability Levels	Factor
1 No information is available.	0
2 The goal achievement report for the year preceding the rating date is prepared by the utility, without verifiable evidence or supporting documentation.	0.5
3 The goal achievement report for the year preceding the rating date is prepared by the utility and includes sufficient verifiable evidence or supporting documentation.	0.9
4 The goal achievement report for the year preceding the rating date is certified by the "board of directors" or by an independent entity.	1

Glossary: Board of directors

Table 61

Reliability Levels	Factor
1 Cannot be corroborated.	0
2 Evidence exists in minutes or summaries of meetings, or in any other medium that certifies it was carried out in at least one of the 3 months or quarters preceding the rating date, depending on whether it is the chief executive office or the "board of directors", respectively.	0.6
3 Evidence exists in minutes or summaries of meetings, or in any other medium that certifies it was carried out in at least 2 of the 3 months or quarters preceding the rating date, depending on whether it is the chief executive office or the "board of directors", respectively.	0.9
	Evidence exists in minutes or summaries of meetings, or in any other medium that certifies it was carried out in the 3 months or

| 4 | quarters preceding the rating date, depending on whether it is the chief executive office or the "board of directors", respectively. | 1 |

Glossary: Board of directors

Table 62

	Reliability Levels	Factor
1	A detailed organizational chart of the utility is not available.	0
2	A detailed organizational chart of the utility exists up to unit level, but its consistency cannot be verified with other sources of information.	0.5
3	A detailed organizational chart of the utility exists up to unit level and is consistent in more than 80% with the positions described in the respective handbook (if applicable) and with the position list considered in the human resources information system.	0.8
4	A detailed organizational chart of the utility exists and is completely consistent with the positions described in the respective handbook (if applicable) and with the position list considered in the human resources information system.	1

Table 63

Reliability Levels	Factor
1 No supporting documents exist.	0
2 The contents of the handbook include, at the very least, descriptions of functions, responsibilities and authority.	0.5
3 The handbook includes the description of all aspects indicated in the practice.	1

Table 64

Reliability Levels	Factor
1 No record exists.	0
2 Records exist for the calendar year preceding the rating date, but it is not possible to verify their consistency with the data in the human resources information system.	0.5
3 Quarterly records exist for the calendar year preceding the rating date and are consistent with the data in the human resources information system.	0.7
4 Monthly records exist for the calendar year preceding the rating date and are consistent with the data in the human resources information system.	0.8
5 Quarterly records exist for the 2 calendar years preceding the rating date and are consistent with the data in the human resources information system.	0.9
6 Monthly records exist for the 2 calendar years preceding the rating date and are consistent with the data in the human resources information system.	1

Table 65

Reliability Levels	Factor
1 Cannot be corroborated.	0
2 Curriculum vitae of respective staff.	0.8
3 Curriculum vitae supported by respective certificates.	1

Table 66

Reliability Levels	Factor
1 Cannot be corroborated.	0
2 Evidence exists of its implementation at least once in the 5 calendar years preceding the rating date.	0.7
3 Evidence exists of its implementation at least 2 times in the 5 calendar years preceding the rating date.	0.9
4 Evidence exists of its implementation at least 3 times in the 5 calendar years preceding the rating date.	1

Table 67

Reliability Levels	Factor
1 Cannot be corroborated.	0
2 It is documented, but no evidence exists of its application at the rating date or in the preceding calendar year.	0.5
3 It is documented and evidence exists of its application in the calendar year preceding the rating date.	0.8
4 It is documented and evidence exists of its application at the rating date.	0.9
5 It is documented and evidence exists of its application at the rating date and in the preceding calendar year.	1

Table 68

Reliability Levels	Factor
1 No record exists.	0
2 Only overall records exist indicating the total number of staff recruited competitively.	0.4
3 Detailed records exist of all staff recruited, indicating recruitment date and procedure by which they were recruited, but without providing evidence.	0.8
4 Detailed records exist, including documentary evidence and the results of the competitive recruitment processes carried out.	1

Table 69

Reliability Levels	Factor
1 No record exists.	0
2 Only overall records exist indicating the total number of staff recruited.	0.4
3 Detailed records of staff recruited exist, but no supporting documents are available.	0.8
4 Detailed records of staff recruited exist and are supported by entries in the human resources information system.	1

Table 70

Reliability Levels	Factor
1 No record exists.	0
2 Only overall records exist indicating the total number of staff participating in "training courses".	0.4
3 Detailed records exist of staff participating in each "training course" and of the number of hours of each course, but no supporting documents are available.	0.8
4 Detailed records of participation in "training courses" exist and are supported by records or accreditation certificates.	1

Glossary: Training courses

Table 71

Reliability Levels	Factor
1 No record exists.	0
2 An annual report exists of the number of utility staff and of the number of staff joining and leaving the entity, classified by type of contract.	0.4
3 A quarterly report exists of the number of utility staff and of the number of staff joining and leaving the entity. It is classified by type of contract (permanent, temporary, other) and is supported by data in the human resources information system.	0.8
4 A monthly report exists of the number of utility staff and of the number of staff joining and leaving the entity. It is classified by type of contract (permanent, temporary, other) and is supported by data in the human resources information system.	1

Table 72

Reliability Levels	Factor
1 No record exists.	0
2 Only overall records exist indicating the total number of staff holding "key positions" who comply with the respective "job descriptions".	0.4
3 Detailed records exist indicating the number of staff holding "key positions" who comply with the respective "job descriptions", but supporting documents are not available.	0.8
4 Detailed records exist and are supported by a study that assesses the degree to which "key position" holders comply with "job descriptions". The study is less than 3 years old and has been updated according to changes in job descriptions and/or in position holders.	1

Glossary: Key positions, Job description

Table 73

Reliability Levels	Factor
1 No record exists.	0
2 Only overall records exist indicating the total number of staff holding "key positions".	0.4
3 Detailed records of "key positions" and of the staff holding them exist, but no supporting documents are available.	0.8
4 Detailed records exist and are supported by a handbook containing "job descriptions".	1

Glossary: Key positions, Job description

Table 74

Reliability Levels	Sum
1 Existence of the system cannot be verified.	0
2 Documents describing the system exist, as do handbooks for its use and maintenance. (Maintenance criteria are not enforceable if the utility uses a system provided by the government).	0.25
3 Staff are trained to use and maintain it.	0.25
4 It is permanently installed on all relevant workstations.	0.25
5 Records exist of its systematic use for purchases of goods and services by tender.	0.25

Table 75

Reliability Levels	Factor
1 No record exists.	0
2 Only overall records exist indicating the total value of purchases made by "public tender"	0.4
3 Detailed records exist of all purchases made indicating their value and the procedure followed (with or without "public tender"), but supporting documents are not available.	0.8
4 Detailed records and supporting documents exist.	1

Glossary: Public tender

Table 76

Reliability Levels	Factor
1 No record exists.	0
2 Only overall records exist indicating the total value of purchases made.	0.4
3 Detailed records exist of all purchases made indicating their value and the procedure followed, but supporting documents are not available.	0.8
4 Detailed records and supporting documents exist.	1

Table 77

	Reliability Levels	Factor
1	No record exists.	0
2	Only overall records exist indicating the total number of "successful" public invitations to tender.	0.4
3	Detailed records exist of all "public" invitations to tender indicating the number of bids, the procedure followed and whether or not they were awarded, but supporting documents are not available.	0.8
4	Detailed records and supporting documents exist.	1

Glossary: Public tender, Successful tenders

Table 78

	Reliability Levels	Factor
1	No record exists.	0
2	Only overall records exist indicating the total number of "public" invitations to tender.	0.4
3	Detailed records exist of all "public" invitations to tender indicating the number of bids, the procedure followed and whether or not they were awarded, but supporting documents are not available.	0.8
4	Detailed records and supporting documents exist.	1

Glossary: Public tender

Table 79

Reliability Levels	Factor
1 No record exists.	0
2 Only overall records exist indicating the total number of invitations to tender that do not exceed the regulated "minimum tender period" by more than 5%.	0.4
3 Detailed records exist of all invitations to tender indicating the timeframe established and the percentage by which they exceeded the "minimum tender period" established in regulations, but supporting documents are not available.	0.8
4 Detailed records exist and are supported by documentary evidence or records of each invitation to tender.	1

Glossary: Minimum tender period

Table 80

Reliability Levels	Factor
1 No record exists.	0
2 Only overall records exist indicating the total number of public invitations to tender without detailing the timeframes within which they were held.	0.4
3 Detailed records exist of all invitations to tender indicating the timeframe established and the percentage by which they exceeded the "minimum tender period" established in regulations, but supporting documents are not available.	0.8
4 Detailed records exist and are supported by documentary evidence or records of each invitation to tender.	1

Glossary: Minimum tender period

Table 81

	Reliability Levels	Factor
1	No record exists.	0
2	A report exists on the number of drinking water and wastewater connections and is supported by an out-of-date register or record (no procedures are in place to update the register or a connection census has not been conducted in the last 2 years) or it cannot be verified.	0.4
3	A quarterly report exists on the number of drinking water and wastewater connections and is supported by an up-to-date register or record (the register or record complies with the requirements of practice FS3.1.2).	0.8
4	A monthly report exists on the number of drinking water and wastewater connections and is supported by an up-to-date register or record (the register or record complies with the requirements of practice FS3.1.2).	1

Table 82

	Reliability Levels	Factor
1	No separate accounting records exist for management and sales expenses or the information comes from financial statements which are incomplete or unaudited, or from audited financial statements receiving either a disclaimer of opinion or an adverse opinion.	0
2	The information comes from financial statements audited by non-"registered" external auditors that include a qualified opinion related to this indicator, or from complementary accounting information "consistent" with the financial statements.	0.3
	The information comes from financial statements audited by "registered" external auditors that include a qualified opinion related to this indicator, or from complementary accounting	

3	information "consistent" with the financial statements.	0.4
4	The information comes from financial statements audited by non-"registered" external auditors that include an unqualified opinion or a qualified opinion not related to this indicator, or from complementary accounting information "consistent" with the financial statements.	0.8
5	The information comes from financial statements audited by "registered" external auditors that include an unqualified opinion or a qualified opinion not related to this indicator, or from complementary accounting information "consistent" with the financial statements.	1

Glossary: Registered external auditors, Consistency of accounting information not originating from the financial statements

Table 83

	Reliability Levels	Factor
1	The practice is not documented.	0
2	A documented procedure exists, but there is no evidence of its application.	0.5
3	A documented procedure exists and "total long-term costs" studies and/or worksheets are available.	0.7
4	A documented procedure exists and applicable tariff studies and/or worksheets are available, including detailed "total long-term costs". The tariff calculation model and its results are reviewed by an external body.	0.8
5	A documented procedure exists and applicable tariff studies and/or worksheets are available, including detailed "total long-term costs". The tariff calculation model and its results are approved by the regulatory body, if applicable, or are audited by an independent external entity.	1

Glossary: Total long-term costs

Table 84

Reliability Levels		Factor
1	The practice is not documented.	0
2	An indexing mechanism is documented, but its application cannot be corroborated.	0.4
3	Evidence exists that the indexing mechanism was applied on the last occasion on which the conditions established for this purpose were met.	0.8
4	Evidence exists that the indexing mechanism was applied on the last 2 occasions on which the conditions established for this purpose were met.	0.95
5	Evidence exists that the indexing mechanism was applied on the last 3 occasions on which the conditions established for this purpose were met.	1

Table 85

Reliability Levels		Factor
1	The practice is not documented.	0
2	Tariffs are differentiated by service type or "system", geographic zone or area served, as appropriate.	0.5
3	Tariffs are differentiated by service type and "system", geographic zone or area served, as appropriate.	1

Glossary: System

Table 86

Reliability Levels	Factor
1 Cannot be corroborated.	0
2 Financial statements for the year or month preceding the rating date are available; depending on whether practice 5 or 6, respectively, is applicable.	0.5
3 Financial statements for the 2 years or months preceding the rating date are available; depending on whether practice 5 or 6, respectively, is applicable.	0.9
4 Financial statements for the 3 years or months, at the very least, preceding the rating date are available; depending on whether practice 5 or 6, respectively, is applicable.	1

Table 87

Reliability Levels	Factor
1 No evidence of financial projections exists.	0
2 Evidence exists, either in documents or spreadsheets, of updated financial projections for a period of 5 or more years, but neither projection bases nor data supporting calculation are included.	0.5
3 Evidence exists, either in documents or spreadsheets, of updated financial projections for a period of 5 or more years and includes projection bases, but does not include either supporting documents or detailed projected quantities and prices for revenue, expenses and cash outflows.	0.6
4 Evidence exists, either in documents or spreadsheets, of updated financial projections for a period of 5 or more years and includes projection bases and supporting documents, but does not include detailed projected quantities and prices for revenue, expenses and cash outflows.	0.7

5 Evidence exists, either in documents or spreadsheets, of updated financial projections for a period of 5 or more years and includes projection bases, supporting documents and detailed projected quantities and prices for revenue, expenses and cash outflows for each of the years projected. | 1

Table 88

	Reliability Levels	Factor
1	The financial statements are incomplete or unaudited, or are audited and include a disclaimer of opinion or an adverse opinion.	0
2	Financial statements audited by non-"registered" external auditors that include a qualified opinion related to this indicator	0.3
3	Financial statements audited by "registered external auditors" that include a qualified opinion related to this indicator	0.4
4	Financial statements audited by non-"registered" external auditors that include an unqualified opinion or a qualified opinion not related to this indicator.	0.8
5	Financial statements audited by "registered external auditors" that include an unqualified opinion or a qualified opinion not related to this indicator.	1

Glossary: Registered external auditors

Table 89

Reliability Levels	Factor
1 The practice is not documented.	0
2 The practice is documented, but no evidence exists of its application.	0.5
3 The practice is documented and evidence exists of its application in the year in which the rating is carried out or in the preceding calendar year.	1

Table 90

Reliability Levels	Factor
1 The practice is not documented.	0
2 A handbook or rules of practice exist, but there is no record of their approval.	0.5
3 Evidence exists of their approval by senior management.	0.8
4 Evidence exists of their approval by the "board of directors".	1

Glossary: Board of directors

Table 93

Reliability Levels		Factor
1	Cannot be corroborated.	0
2	The practice is documented, but no evidence exists of its application.	0.5
3	Evidence exists of its application in the billing period preceding the rating date.	0.8
4	Evidence exists of its application in the 3 billing periods preceding the rating date.	0.9
5	Evidence exists of its application in the 6 billing periods preceding the rating date.	1

Table 94

Reliability Levels		Factor
1	Cannot be corroborated.	0
2	The practice is documented, but no evidence exists of its application.	0.5
3	The practice is documented and evidence exists of its application in the billing period preceding the rating date.	0.7
4	The practice is documented and evidence exists of its application in the 3 billing periods preceding the rating date.	1

Table 95

Reliability Levels	Factor
1 Cannot be corroborated.	0
2 The practice is documented, but no evidence exists of its application.	0.5
3 The practice is documented and evidence exists of its application at the rating date.	1

Table 96

Reliability Levels	Factor
1 Cannot be corroborated.	0
2 The practice is documented and evidence exists of its application at the rating date.	1

Table 97

	Reliability Levels	Factor
1	Cannot be corroborated.	0
2	The practice is documented, but no evidence exists of its application.	0.5
3	Evidence exists of application of the policy to at least one of the forms of fraud in the month preceding the rating date. Moreover, in the case that estimated losses attributable to users surpassed 10% of unbilled water volume, evidence exists that a procedure to detect illegal connections was carried out in the rating year or that 2 were carried out the previous year.	0.7
4	Evidence exists of application of the policy to at least 2 of the forms of fraud in the month preceding the rating date. Moreover, in the case that estimated losses attributable to users surpassed 10% of unbilled water volume, evidence exists that 2 procedures to detect illegal connections were carried out in the rating year or that 3 were carried out the previous year.	0.9
5	Evidence exists of comprehensive application of the fraud detection policy in the month preceding the rating date. Moreover, in the case that estimated losses attributable to users surpassed 10% of unbilled water volume, evidence exists that 3 procedures to detect illegal connections were carried out in the rating year or that 4 were carried out the previous year.	1

Table 98

Reliability Levels	Factor
1 No record exists.	0
2 Paper records.	0.5
3 Computerized records not integrated with the accounting system.	0.8
4 Computerized records integrated with the accounting system.	1

Table 99

Reliability Levels	Factor
1 The financial statements are incomplete or unaudited, or are audited and include a disclaimer of opinion or an adverse opinion.	0
2 Financial statements audited by non-"registered" external auditors that include a qualified opinion related to this indicator, or complementary accounting information "consistent" with the financial statements.	0.3
3 Financial statements audited by "registered external auditors" that include a qualified opinion related to this indicator, or complementary accounting information "consistent" with the financial statements.	0.4
4 Financial statements audited by non-"registered" external auditors that include an unqualified opinion or a qualified opinion not related to this indicator, or complementary accounting information "consistent" with the financial statements.	0.8
5 Financial statements audited by "registered external auditors" that include an unqualified opinion or a qualified opinion not related to this indicator, or complementary accounting information "consistent" with the financial statements.	1

Glossary: Registered external auditors, Consistency of accounting information not originating from the financial statements

Table 100

Reliability Levels	Factor
1 Estimate without sufficient substantiation.	0
2 The estimated number of inhabitants is based on data from a "property" or "user" register, without evidence of that register being updated in the three years preceding the rating date.	0.5
3 The estimated number of inhabitants is based on data from a "property" or "user" register and evidence exists of that register being updated in the three years preceding the rating date, and the data are compared with records of connections in the "geographical area to be rated".	0.75
4 The estimated number of inhabitants is based on data from a "property" or "user" register and evidence exists of that register being updated in the year preceding the rating date, and the estimate is based on a ratio of inhabitants per dwelling and is supported by data published by a "competent official body", and the data are compared with records of connections in the "geographical area to be rated". Otherwise, the number of inhabitants is taken from an estimate for the calendar year being rated published by a "competent official body".	1

Glossary: Property, Active users, Geographical area to be rated, Competent official body

Table 101

Reliability Levels	Factor
1 Data without sufficient substantiation.	0
2 Data are taken from an estimate prepared by the utility or by a third party and the estimate is "substantiated" by data published by a "competent official body".	0.5
3 Data are taken from an estimate published by a "competent official body".	1

Glossary: Competent official body, Substantiated estimate

Table 103

Reliability Levels	Factor
1 Sufficient documentation does not exist.	0
2 The corresponding tariff, applicable amount of subsidy (if appropriate) and expense calculation are duly documented.	1

Table 104

Reliability Levels	Factor
1 Data without sufficient substantiation.	0
2 Data are taken from a statistic published by a "competent official body", but they refer to a period more than 5 years before the calendar year being rated.	0.5
3 Data are taken from a statistic published by a "competent official body", they refer to a geographical area that is identical or similar to the "geographical area to be rated", and they refer to a period between 2 and 5 years before the calendar year being rated.	0.9
4 Data are taken from a statistic published by a "competent official body", they refer to a geographical area that is identical or similar to the "geographical area to be rated", and they refer to a period less than 2 years before the calendar year being rated.	1

Glossary: Geographical area to be rated, Competent official body

Table 105

Reliability Levels	Factor
1 No supporting instruments or documents exist.	0
2 Applicable supporting instruments or documents, duly approved by the competent bodies, exist.	1

Table 106

Reliability Levels	Factor
1 No currently applicable supporting documents exist.	0
2 Currently applicable supporting documents exist, but there is no evidence in writing of their application.	0.5
3 Currently applicable supporting documents exist and there is evidence in writing of their application at the rating date or on the last corresponding occasion.	1

Table 107

Reliability Levels	Factor
1 No currently applicable supporting documents exist.	0
2 Currently applicable supporting documents exist, but there is no evidence in writing of their application.	0.5
3 Currently applicable supporting documents exist and there is evidence in writing of their application at the rating date or in one of the 2 calendar years preceding that date.	1

Table 108

Reliability Levels	Factor
1	No evidence exists of "publication on the institutional website" or in other media.	0
2	Evidence exists of "publication on the institutional website" or in traditional media at the rating date, but the information is out of date.	0.2
3	Evidence exists of "publication" of up-to-date information on the institutional website or in traditional media at the rating date, but the information published is not approved by the competent external authority/internal unit responsible (as appropriate).	0.4
4	Evidence exists of publication of up-to-date information in traditional media at the rating date, and the information published is approved by the competent external authority/internal unit responsible (as appropriate).	0.9
5	Evidence exists of "publication" of up-to-date information on the institutional website at the rating date, and the information published is approved by the competent external authority/internal unit responsible (as appropriate).	1

Glossary: Publication on the institutional website

Table 109

Reliability Levels	Factor
1 Estimated volumes withdrawn	0
2 Monthly records exist of volumes either withdrawn from the environment or obtained from another system's infrastructure and incorporated into the supply, treatment or distribution "system".	0.7
3 Daily records exist for more than 95% of volumes either withdrawn from the environment or obtained from another system's infrastructure and incorporated into the supply, treatment or distribution "system".	0.9
4 Daily records, produced by remotely controlled gauging systems, exist of all volumes either withdrawn from the environment or obtained from another system's infrastructure and incorporated into the supply, treatment or distribution "system".	1

Glossary: System

Table 110

Reliability Levels	Factor
1 No records or gauging data exist for estimated inflows to "water bodies".	0
2 Partial circulating flow and/or piezometer records exist (per gauging point or short or incomplete time series). Data series are shorter than 5 years or contain errors or are missing data for more than 30% of the days.	0.5
3 Circulating flow and/or piezometer records exist for official, well-maintained gauging networks in all bodies and at all available intake points. At least 5 years' complete data series are available.	1

Glossary: Water bodies

Table 111

Reliability Levels	Factor
1 Estimated energy consumption.	0
2 Partial or insufficiently long (less than 3 years) records of energy consumption exist.	0.33
3 Partial or insufficiently long (less than 3 years) records of energy consumption exist and are produced by calibrated metering devices or supported by accrediting documents issued by energy supply companies.	0.67
4 Energy consumption records exist for all points of consumption and are produced by calibrated metering devices or supported by accrediting documents issued by energy supply companies.	1

Table 112

Reliability Levels	Factor
1 Estimated energy generation.	0
2 Partial or insufficiently long (less than the last 3 full years) energy generation records exist.	0.33
3 Energy generation records exist for part of the facilities or for insufficiently long time series (less than the last 3 full years) and are produced by metering devices or supported by accrediting documents issued by energy-buying companies.	0.67
4 Energy generation records exist for all the facilities and are produced by metering devices or supported by accrediting documents issued by energy-buying companies.	1

Table 113

Reliability Levels		Factor
1	Estimated emissions and general ratios.	0
2	Records and meter readings exist of the utility's overall energy consumption, as do estimates of its transformation into direct emissions and estimates of vehicle and machinery emissions. Explicit criteria exist to determine equivalence with other gases.	0.8
3	Records and meter readings exist of electricity consumption in all operational centers and in all other facilities that generate direct emissions, such as heating boilers, and estimates exist of direct emissions by vehicles and machinery. Explicit criteria exist to determine equivalence with other gases.	1

Table 114

Reliability Levels		Factor
1	No transportation or destination records exist.	0
2	Records of transportation and destination exist for part of the sludge generated.	0.5
3	Paper records of transportation and destination exist for all sludge generated.	0.8
4	Computerized records of transportation and destination exist for all sludge generated.	1

Table 115

Reliability Levels	Factor
1 No records of the sludge generated exist.	0
2 Records exist for part of the sludge generated.	0.5
3 Paper records exist for all sludge generated.	0.8
4 Computerized records exist for the sludge generated in each facility.	1

Table 116

Reliability Levels	Factor
1 Either no records exist or they are based on estimates.	0
2 Paper records exist for wastewater and wastewater treatment infrastructure on land registry and commercial maps and estimates exist about the census population's geographical distribution. An explicit criterion exists for calculating discharge in terms of "population equivalent".	0.8
3 Paper records exist for wastewater and wastewater treatment infrastructure on land registry and commercial maps and estimates exist about the census population's geographical distribution. An explicit criterion exists for calculating discharge in terms of "population equivalent".	0.9
4 Records exist for wastewater and wastewater treatment infrastructure on land registry and commercial maps in a GIS system and all wastewater collection network connections are mapped according to accurate census data. An explicit criterion exists for calculating discharge in terms of "population equivalent".	1

Glossary: Population equivalent

Table 117

Reliability Levels	Factor
1 Data without sufficient substantiation.	0
2 Censuses exist of inhabitants, properties and commercial and industrial activities, or estimates exist of the number of inhabitants, properties and commercial and industrial activities in the "geographical area to be rated". These data, published by a "competent official body", refer to a period more than 5 years before the calendar year rated. An explicit criterion exists for calculating discharge in terms of "population equivalent".	0.5
3 Censuses exist of inhabitants, properties and commercial and industrial activities, or estimates exist of the number of inhabitants, properties and commercial and industrial activities in the "geographical area to be rated". These data, published by a "competent official body", refer to a period less than 5 years and more than 1 year before the calendar year rated. An explicit criterion exists for calculating discharge in terms of "population equivalent".	0.8
4 Censuses exist of inhabitants, properties and commercial and industrial activities, or estimates exist of the number of inhabitants, properties and commercial and industrial activities in the "geographical area to be rated". These data, published by a "competent official body", refer to a period less than 1 year before the calendar year rated. An explicit criterion exists for calculating discharge in terms of "population equivalent".	1

Glossary: Population equivalent, Geographical area to be rated, Competent official body

Table 118

Reliability Levels	Factor
1	No record exists.	0
2	Measurements are documented in unsigned records with no quality control.	0.5
3	Measurements are documented in signed paper records subject to traceability criteria and quality control.	0.9
4	Measurements are documented in signed computerized records subject to traceability criteria and quality control.	1

Table 119

Reliability Levels	Factor
1	Neither an inventory of monitoring points nor records of investigations or reports exist.	0
2	A system exists to record and track environmental investigations.	0.33
3	A system exists to monitor and self-assess compliance with environmental regulations.	0.67
4	An inventory exists of "points potentially monitored" for the reasons stated in this indicator and records of reports, investigations, etc. exist in a quality system, computerized system or equivalent.	1

Glossary: Monitoring points

Table 120

Reliability Levels		Factor
1	The practice is not documented.	0
2	The practice is documented, but there is no evidence of its application even though it should have been applied in the 3 calendar years preceding the rating date.	0.5
3	The practice is documented and evidence exists of its application at the rating date or in the calendar year preceding such date, if applicable.	0.7
4	The practice is documented and evidence exists of its application at the rating date and in the 2 calendar years preceding such date, if applicable. Or, the practice was not applicable in that period.	0.9
5	The practice is documented and evidence exists of its application at the rating date and in the 3 calendar years preceding such date, if applicable. Or, the practice was not applicable in that period.	1

Table 121

Reliability Levels		Factor
1	The practice is not documented.	0
2	The practice is documented, but there is no evidence of its application in the calendar year preceding the rating date.	0.3
3	The practice is documented and evidence exists of its application in the calendar year preceding the rating date, including records of public tap locations and records of quality control performed on the water supplied.	0.8
4	The practice is documented and evidence exists of its application in the 2 calendar years preceding the rating date, including records of public tap locations with references to zones served and records of quality control performed on the water supplied.	1

Table 122

Reliability Levels	Factor
1	Paper records exist of malfunctions and of processes and work orders up to resolution and user notification, if applicable. Classified by type and scope of disruption.	0
2	Computerized records exist of "malfunctions" and of processes and work orders up to resolution and user notification, if applicable. Classified by type and scope of disruption.	0.5
3	Computerized records exist of "malfunctions" and of processes and work orders up to resolution and user notification, if applicable. Classified by type and scope of disruption.	0.7
4	Computerized records exist (in geo-referenced databases linked to user management systems) of "malfunctions" and of processes and work orders up to resolution and user notification, if applicable. Classified by type and scope of disruption.	1

Glossary: Malfunction

Table 1088

Reliability Levels	Factor
1	i) Does not submit balance sheet or income statement at the "rating scope" level, or does not submit information regarding the financial position and economic income by business unit, or information submitted is "not consistent" with the financial statements at the utility level or is audited and includes either a disclaimer of opinion or an adverse opinion; or ii) The financial statements of the utility are incomplete or unaudited, or are audited and include either a disclaimer of opinion or an adverse opinion.	0
2	i) The unaudited balance sheet and income statement at "rating scope" level are "consistent" and "reconciled" with the utility's financial statements audited by "registered" or non-registered external auditors and include a qualified opinion related to this indicator; or ii) The financial statements at "rating scope" level are audited by registered or non-registered external auditors and include a qualified opinion related to this indicator; or iii) The financial position and economic income at the "rating scope" level and of all other business units are presented in an itemized way in a note to the utility's financial statements and are "reconciled" with the data contained in such statements audited by registered or non-registered external auditors and include a qualified opinion related to this indicator.	0.4
3	The unaudited balance sheet and income statement at "rating scope" level are "consistent" and "reconciled" with the utility's financial statements audited by non-"registered" external auditors and include a qualified opinion not related to this indicator and are published in the institution's report or equivalent document.	0.6
4	The unaudited balance sheet and income statement at "rating scope" level are "consistent" and "reconciled" with the utility's financial statements audited by "registered external auditors" and include an unqualified opinion or a qualified opinion not related to this indicator and are published in the institution's report or equivalent document.	0.7

i) The financial statements at rating scope level and utility level are audited by non-"registered" external auditors and include an unqualified opinion or a qualified opinion not related to this indicator.

5   ii) The financial position and economic income at the "rating scope" level and of all other business units are presented in an itemized way in a note to the utility's financial statements and are "reconciled" with the data contained in such statements audited by non-registered external auditors and include an unqualified opinion or a qualified opinion not related to this indicator.   0.8

i) The financial statements at "rating scope" level and utility level are audited by "registered external auditors" and include an unqualified opinion or a qualified opinion not related to this indicator.

6   ii) The financial position and economic income at the "rating scope" level and of all other business units are presented in an itemized way in a note to the utility's financial statements and are reconciled with the data contained in such statements audited by "registered external auditors" and include an unqualified opinion or a qualified opinion not related to this indicator.   1

Glossary: Registered external auditors, Rating scope, Consistency of financial statements at rating scope level, Reconciled data

Appendix B - Glossary Terms

Acceptable timeframe for circulation of minutes issued by the board
Not more than 15 working days after the meeting. In the case that decisions made and recorded in the minutes are relevant to the next meeting: not less than 2 working days before the respective meeting.

Acceptable timeframe for circulation of minutes issued by the body that represents the owners
Not more than 15 working days after the meeting. In the case that decisions made and recorded in the minutes are relevant to the next meeting: not less than 2 working days before the respective meeting.

Access
Access on the plot or in the dwelling to a minimum guaranteed volume of over 40 liters of drinking water per person per day. Households provided with a domestic connection but which are served with water that cannot be considered drinkable or which are not delivered the indicated minimum amount will not be counted as having access in this section.

Active connections
Individuals physically connected to the drinking water and/or sewerage network managed by the utility and receiving the service. Connections must have at least one user and may have several.

Active users
Users registered with the utility to use or consume drinking water and/or wastewater services and receiving those services.

Applicable regulations

Legislation and standards with which compliance is obligatory within the geographical service coverage area.

Appropriate cost of capital
Rate established by the regulating authority, or, if unregulated: i) rate determined according to the CAPM (Capital Asset Pricing Model), which considers that an asset's rate of return depends on the risk-free rate plus a risk premium, including systematic and unsystematic risk; or ii) the WACC (Weighted Average Cost of Capital), which corresponds to the average of the cost of debt and the opportunity cost of capital, weighted on the basis of their contribution to the financial structure.

Board of directors
(also called: the board)
The board of directors is a necessary and permanent affiliate body whose members are designated at regular intervals by the meeting of stockholders or by the body that represents the owners. Its function is to carry out all ordinary and extraordinary acts of administration, representing the utility before third parties and accepting joint liability for breach of the duties assumed under law, regulations, articles of incorporation and bylaws.

Body that represents the owners
In corporations (public or private), this body is the meeting of stockholders (or general meeting), while in other limited companies it is the meeting of owners. In cooperatives, this body is the meeting of cooperative members. In public companies devolved from the national or sub-national state, this body is typically the head of the executive body (and to some effects the head of the legislative body).

Business autonomy (for decisions related to remunerations and staffing, to acquisition of goods and services, and to debt)
The utility has business autonomy in all cases in which ordinary and extraordinary acts associated with the above-mentioned activities are governed by the laws applicable to private companies in the respective country (e.g. labor law, company law, etc.).

Competent official body
Authority that has the power, among those granted by the state, to determine, publish or verify the number of inhabitants and/or dwellings and/or household income or expenditure and/or population data in general and/or the coverage of water and/or wastewater services in a territory within the utility's geographical area to be rated.

Complaint

Notification by a user/customer of a perceived anomaly in any of the service processes, including those linked to customer management.

Consistency of accounting information not originating from the financial statements

Accounting data from sources other than the utility's financial statements are considered consistent with the financial statements if one of the following conditions is satisfied: i) the total amount contained in the ancillary records from which the information originated matches the respective amount in the financial statements; ii) the amount of the item used to calculate the indicator is taken from an account which in turn is contained in any of the items which form part of the financial statements; or iii) the sum of the amounts of the various areas or cost centers into which the items are disaggregated matches the respective amount in the financial statements.

When applying this definition at rating scope level, the following will be considered:
i) The reference made in the table to audited financial statements refers to company level. However, if in addition audited financial statements at rating scope level are submitted, the external auditors' opinion regarding the financial statements at both levels will be considered, applying the most restrictive provided that it is pertinent.
ii) Values not originating from the financial statements will be consistent with both the financial statements at rating scope level and the financial statements at company level.

Consistency of financial statements at rating scope level

Financial statements at rating scope level are considered to be consistent with the utility's financial statements if the horizontal sum, line by line, of the components of the financial statements for the various areas into which the utility's accounts are divided matches the amounts of the respective financial statement's components at utility level once deductions are made for possible transactions between areas, if these have been entered into the accounts.

Contingency

Circumstance that has modified the conditions and factors that frame the service and that may disrupt service quality or continuity.

Conventional water treatment

Set of processes that turn raw water into drinking water and that, in addition to disinfection and filtration, includes some further process.

Corrective maintenance

Set of inspection and repair actions carried out as a consequence of a malfunction, anomaly or incident detected or reported during operation or regular service delivery.

Corrective maintenance protocol
Written specification of the checks and tasks, detailing the persons responsible for carrying them out (how, when, where and who), performed to correct anomalies or malfunctions.

Costs per activity
Methodology that measures performance of activities, resources and cost objects. Resources are assigned first to activities and then activities are assigned to cost objects according to use. Cost objects correspond to the recipient of the work and can be internal (products, services, departments) or external (customers, suppliers).

Criteria established in the International Accounting Standards (IAS 16)
Paragraph 7 of IAS 16 (Recognition of property, plant and equipment).
The cost of an item of property, plant and equipment will be recognized as an asset if, and only if: (a) it is probable that future economic benefits associated with the item will flow to the entity; and (b) the cost of the item can be measured reliably.

Subsequent costs related to property, plant and equipment:
Paragraph 12 of IAS 16 (Maintenance expense of property, plant and equipment).
Under the recognition principle in paragraph 7, an entity does not recognize in the carrying amount of an item of property, plant and equipment the costs of the day-to-day servicing of the item. Rather, these costs are recognized in profit or loss as incurred. Costs of day-to-day servicing are primarily the costs of labor and consumables, and may include the cost of small parts. The purpose of these expenditures is often described as for the 'repairs and maintenance' of the item of property, plant and equipment.

Paragraph 13 of IAS 16 (Investment in replacement of property, plant and equipment).
Parts of some items of property, plant and equipment may require replacement at regular intervals. For example, a furnace may require relining after a specified number of hours of use, or aircraft interiors such as seats and galleys may require replacement several times during the life of the airframe. Items of property, plant and equipment may also be acquired to make a less frequently recurring replacement, such as replacing the interior walls of a building, or to make a non-recurring replacement. Under the recognition principle in paragraph 7, an entity recognizes in the carrying amount of an item of property, plant and equipment the cost of replacing part of such an

item when that cost is incurred if the recognition criteria are met. The carrying amount of those parts that are replaced is derecognized in accordance with the derecognition provisions of this Standard.

Customer service complaint
Notification, complaint or request referring to alleged malfunction of customer service processes and not linked to availability of water supply or wastewater collection services.

Discount rate applicable to capital cost
Rate established by the Regulating Authority, or, if unregulated: i) rate determined according to the CAPM (Capital Asset Pricing Model), which considers that an asset's rate of return depends on the risk-free rate plus a risk premium, including systematic and unsystematic risk; or ii) the WACC (Weighted Average Cost of Capital), which corresponds to the average of the cost of debt and the opportunity cost of capital, weighted on the basis of their contribution to the financial structure.

Emergency
Unforeseen or unplanned event of unknown origin that may significantly degrade the quantity or quality of water and/or wastewater services delivered to users. It may be minor and local or major and widespread and may be attributable to natural causes (earthquake, hurricane, flood, forest fire, drought, freeze, etc.) or human action (error, transport accident, vandalism, civil unrest, terrorism, etc.).

Entry point into the drinking water supply system
Refers to each point in the drinking water supply infrastructure able to incorporate water into the system through intakes of surface or groundwater, or through connections to other systems from which water may be imported. Connections to other systems may allow export of water from the rated system, rendering negative flow values. Water may be raw or treated according to the network point at which it is incorporated and the conditions of the water taken in or imported.

Environmentally responsible destination
Any destination that complies with or improves upon applicable regulations.

Financial expenses
These include expenses for interest on resources obtained from third parties to finance utility operation and investment, such as interest on bank loans, interest on loans from multilateral financial institutions and interest on public debt instruments.

Finished works
Works are considered finished as of the date of their start-up, as authorized by the utility.

Formal autonomy (to acquire goods and services and make payments, to set remunerations and determine staffing; and to contract debt).
The utility has formal autonomy in all cases in which ordinary or extraordinary acts associated with the above-mentioned activities do not require authorization from an external body.

Geographical area to be rated
Corresponds to the geographical area in which the utility is responsible for delivering water and/or wastewater services, as defined in the utility's mandate (or mandates) to provide the services to be rated.

Geographical service coverage area
Local geographical area in which a utility has a legal or contractual duty to provide a service.

In force (investment plan in force)
A plan is considered to be in force as of the date of its approval by the utility's board of directors.

Incident
Alteration to the characteristics or operating conditions of a system component that has modified the conditions of the service delivered.

Independent director
Does not have significant financial or political ties to the management, to the majority owner(s), to the head of the state body that acts as owner in the case of public companies, or to other parts of the company that may interfere with independent exercise of the director's judgment in the sole interests of the company.

Integrated project monitoring system
Information system that contains the history of costs and time periods of each project throughout its life cycle.

Job description
Set of general and specific characteristics a person must possess in order to hold a position. For the purposes of measuring this indicator, the institution must have specified, at the very least, the academic qualifications, experience (general and specific) and technical competences required to perform the job.

Key positions
For this purpose, key positions are considered to be all supervisory positions and all those in charge of plants, laboratories and critical equipment.

Legal person(ality)
Entity (other than a natural person) with the capacity to acquire rights and contract obligations.

Malfunction
Anomalous functioning of a system component.

Management control system
Set of processes used to analyze data and information (not necessarily computerized) which, by regularly and systematically generating indicators, indices and tables and graphs, allows managers to monitor the organization's performance in order to achieve the desired results.

Mandate
Authority of a utility to deliver certain drinking water and/or wastewater services in a defined geographical area. The mandate is given to the utility by law according to its institutional nature, through entrustment, or by contract awarded by an institution with the legal power to do so.

Minimum tender period
Refers to the sum of the minimum time periods established in the respective regulation for each phase of the tender (from publishing the request for proposals to notifying the award of the tender) which has to be observed to allow for bidder participation and appropriate evaluation of proposals.

Monitoring points
Points on the watershed's surface and groundwater networks at which environmental regulations establish conditions determining environmental flow or quantitative or qualitative characteristics for discharge of wastewater, sludge or other air or noise pollutants. Waste transport will be considered as a single point of interest.

Operational expenses
These include expenditure on staff benefits, raw materials and consumables, depreciation and amortization and other operational expenses (if expenses are classified by nature in the financial statements). If expenses are classified by function, these include operating costs, management expenses and distribution or sales expenses, including depreciation and amortization for the financial period.

Own staff
Includes all staff directly hired by the utility in any legal form (employee, contractor, permanent worker, temporary worker, etc.). For part-time staff, full-time equivalents must be established. (If corporate staff work in organizations that include other businesses apart from drinking water and/or wastewater, said staff must be distributed in proportion to the number of workers in each business when calculating the productivity indicator).

Owner (utility)
Natural or legal person (private or public) exercising owner rights over the utility (rights of disposal, use and usufruct, among others). The term 'owner' generally applies to organizations or companies governed by private law. It is a constituent characteristic of companies that have one or more owners exercising associate owner rights. For organizations governed by public law (e.g. a department in the executive arm of a government), the term 'owner' is not appropriate when referring to the entity that exerts control over the utility and therefore should be replaced by 'competent state representative'. In republics/states governed by rule of law it is ultimately sovereign citizens who delegate administration of organizations governed by public law to the state's representatives and who delegate to them the power to exercise owner rights over publicly owned companies.

Population equivalent
Population calculated by adding to the census population the number of inhabitants equivalent to the volumes and quality used by commerce and industry in the area rated.
It is a regular practice in evaluating liquid loads discharged into wastewater treatment plants.

Preventive maintenance
Set of inspection and repair actions carried out according to a program and in anticipation of signs or report of anomalies or malfunctions.

Preventive maintenance protocol
Written specification of the checks, actions and tasks, detailing the persons responsible for carrying them out (how, when, where and who), performed to ensure operational capacity of equipment and facilities, and, if necessary, to initiate pertinent corrective measures.

Property
Dwellings, shops or businesses that receive the service and which are individually identified in residential, industrial or commercial censuses, or in the utility's records and databases.

Public debt instruments
Any instrument that represents debt or credit (such as bonds, promissory notes, etc.) acquired via public offering.

Public tender
Procedure under which a public call for proposals is issued via public access media and the most suitable proposal received is accepted.

Publication on the institutional website
The information is understood to be published when it is accessible from the utility's institutional website, either directly or via a reference and/or a link to a competent body's website which allows users to access the information.

Rating scope
Scope of application of the AquaRating, defined jointly by the scope of services to be rated and the geographical area to be rated.

Real time
For this document's purposes, real time is defined as data transmission with a time lag shorter than a minute.

Reclaimed water
Water that after being used and collected by sewer networks is treated to achieve compliance with certain quality standards, making it suitable for use in specified activities.

Reconciled data
Information on the financial position and economic income at rating scope or business segment level is considered reconciled with the contents of the financial statements at company level if that information is presented in a table which demonstrates fulfilment of the criteria of consistency of financial statements at rating scope level.

Recycled/Recycling
These terms usually refer to water reuse within the private domain at dwelling, business or condominium level.

Registered external auditors
Auditors registered with a multilateral financial institution or with the respective regulatory body responsible for overseeing corporations or equivalent public bodies. In the case of small companies (fewer than 10,000 water connections), the statutory auditor established in some national legislation (e.g. in Colombia) is considered equivalent to registered external auditors.

Registered user
Service user individually identified in any of the records or databases.

Representative sample of supplied quality
Sample corresponding to each zone served by a water treatment plant (or by a source that does not require treatment) with a population equal to or less than 10,000 inhabitants. In served zones with populations of over 10,000 inhabitants, a sample will be considered representative when taken per 10,000 served inhabitants with an equivalent geographical distribution.

Return on equity
Amount of income for the financial year as a percentage of equity at the beginning of the financial year, discounting inflation. The rate of inflation is measured by each respective country's most representative price index.

Reused water
Reclaimed water used for the purposes specified by compliance with the quality and monitoring standards established in the applicable regulations.

Satisfied user
User who has expressed satisfaction (defined as a value above the satisfaction threshold) in consultations or surveys carried out by the utility or competent authority.

Scope of services to be rated
Corresponds to drinking water and/or wastewater service delivery, as defined by the set of all service stages and service functions included in the utility's mandate (or mandates) to provide the services to be rated.

Service function(s)
AquaRating considers the following four service functions: Customer management; operation and maintenance of infrastructure linked to service stages implemented; financing of replacement of existing physical assets; and financing of augmentation of existing assets or of new physical assets.

Service stage
AquaRating defines the following service stages: Drinking water production, including intake and treatment; drinking water distribution; wastewater collection; and final wastewater disposal with or without treatment (of any type), including marine and river outfall and any other discharge.

Strategic map
Visual representation of an organization's strategy. It provides an overview and summarizes strategic essentials to facilitate communication and

implementation. The strategic map depicts the cause-effect relationships between the various objectives or perspectives considered in the strategic plan.

Substantiated estimate
Estimate of the number of inhabitants, prepared by the utility or by a third party, based on population data taken from a population and/or dwelling census published by a competent official body. Data used in the estimate will include at least the following: Inhabitants of the geographical area to be rated in a base year selected within the previous 10 years, or ratio of inhabitants per dwelling for the geographical area to be rated in a base year selected within the previous 10 years.

Successful tenders
Public tenders for which 3 or more proposals are received, one of which is awarded the contract, are considered successful. Tenders that have to be published again due to failure of the first one, even though they receive 3 or more offers and one of those is awarded the contract, are not considered successful.

Sufficient hydraulic conditions for use and consumption
Pressure, water quality and flow parameters that enable water consumption at each connection under the contracted terms or under the general terms established by the applicable regulations.

Supply-demand balance
Refers to comparative water supply and demand at a specific point within a system, usually at a water treatment plant's output. The concept is also applied to balances in specific facilities and may include annual average daily flow, peak daily flow or peak hourly flow, as appropriate.

System
Set of infrastructure, facilities and equipment used or available to deliver water and/or wastewater services in part or all of a geographical service coverage area and which are linked topologically and hydraulically. A system will always be related to a servable geographical area.

Technical expertise
Skills and knowledge needed to perform a function.

Timely reception of notices (board members)
The following are considered 'timely reception': for notices that set meeting dates, at least 5 working days; for meeting agendas, at least 2 working days.

Timely reception of notices (members of body that represents the owners)
The following are considered 'timely reception': for notices that set meeting dates, at least 10 working days; for meeting agendas, at least 5 working days.

Total long-term costs
Cost of investment to augment and replace existing assets and optimize operation and maintenance to satisfy long-term service demand (time horizon of at least 15 years).

Training courses
For this purpose, only courses offering at least 4 hours' training are considered.

Wastewater treatment plan
Duly approved document which contains a program of investment in (and performance of, if applicable) any type of wastewater treatment before returning the water to the natural environment within the geographical area to be rated.

Water bodies
Water bodies are understood to be the parts of the watershed that share homogeneous characteristics in terms of flow regime and biological, physical and chemical quality.

Water volume incorporated into the system
Total volume of water incorporated into the supply and distribution system, whether taken in from the environment or imported from other systems.

Weighted average tax rate
SUM $(I1/I*t1+I2/I*t2...+In/I*tn)$, where $I1$, $I2...In$ corresponds to the part of income from provision of services levied with the respective sales tax, including the tax-exempt part if applicable (sum of $I1+I2...+In = I$); I corresponds to total income from provision of services in the respective period (variable FS3-V12); $t1$, $t2...tn$ corresponds to each tax rate expressed as fractions of 1 (in the case of tax-exempt income the tax rate is 0).

Works (in relation to investment plan projects)
Investment plan projects are usually subdivided into works, which are put out to public tender. Several projects may be grouped into a single works contract for tendering purposes.

Zero-value bill
Bill issued without including charges for service delivery.

Zones at risk of non-compliance with drinking water quality standards
Zones within the geographical service coverage area in which the probability of not complying with the values established by regulations is higher than the threshold set for its users. In the absence of reference values or thresholds for either probability or affected area, probability will be calculated based on the non-compliance records for that area in the 2 years preceding the rating.

Appendix C - AquaRating Weights

Assessment areas weights:

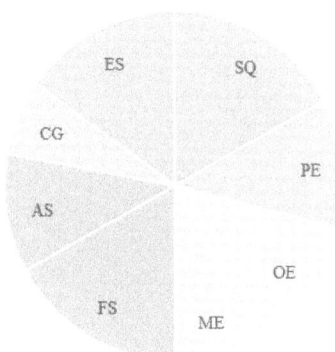

Relative Weight	Absolute Weight	Area	
17	17	SQ	Service Quality
12	12	PE	Investment Planning and Implementation Efficiency
13	13	OE	Operating Efficiency
8	8	ME	Business Management Efficiency
17	17	FS	Financial Sustainability
11	11	AS	Access to Service
7	7	CG	Corporate Governance
15	15	ES	Environmental Sustainability

SQ Service Quality

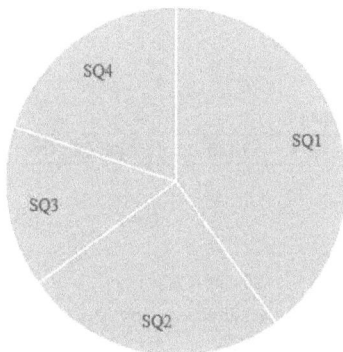

Relative Weight	Absolute Weight	Sub-Area	
40	6,8	SQ1	Drinking water quality
25	4,25	SQ2	Distribution of drinking water for use and consumption
15	2,55	SQ3	Wastewater collection
20	3,4	SQ4	User service

SQ1 Drinking water quality

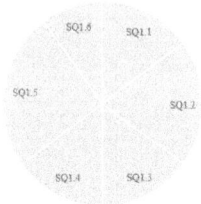

Relative Weight	Absolute Weight	Indicator	
15	1,02	SQ1.1	Assurance of structural capacity for treatment and supply
33,33	0,34	P1	
22,22	0,2267	P2	
11,11	0,1133	P3	
11,11	0,1133	P4	
11,11	0,1133	P5	
11,11	0,1133	P6	
20	1,36	SQ1.2	Assurance of appropriate supplied water quality
9,09	0,1236	P1	
9,09	0,1236	P2	
9,09	0,1236	P3	
27,27	0,3709	P4	
18,18	0,2473	P5	
9,09	0,1236	P6	
9,09	0,1236	P7	
9,09	0,1236	P8	
15	1,02	SQ1.3	Supervision and control of supplied water quality
7,14	0,0729	P1	
14,29	0,1457	P2	
14,29	0,1457	P3	
14,29	0,1457	P4	
21,43	0,2186	P5	
7,14	0,0729	P6	
7,14	0,0729	P7	
14,29	0,1457	P8	
15	1,02	SQ1.4	Structural operational capacity for drinking water treatment
25	1,7	SQ1.5	Compliance with drinking water standards
10	0,68	SQ1.6	Supplied water quality control frequency

SQ2 Distribution of drinking water for use and consumption

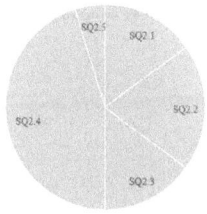

Relative Weight	Absolute Weight	Indicator	
15	0,64	SQ2.1	Assurance of structural capacity for supply and distribution
37,5	0,2391		P1
25	0,1594		P2
25	0,1594		P3
12,5	0,0797		P4
20	0,85	SQ2.2	Assurance of supply continuity during operation
11,11	0,0944		P1
11,11	0,0944		P2
22,22	0,1889		P3
22,22	0,1889		P4
22,22	0,1889		P5
11,11	0,0944		P6
15	0,64	SQ2.3	Supervision and control of supply continuity
33,33	0,2125		P1
33,33	0,2125		P2
33,33	0,2125		P3
45	1,91	SQ2.4	Supply continuity
5	0,21	SQ2.5	Time taken to connect new users to the drinking water service

SQ3 Wastewater collection

Relative Weight	Absolute Weight	Indicator	
17	0,43	SQ3.1	Assurance of structural capacity for wastewater collection
25	0,1084		P1
25	0,1084		P2
25	0,1084		P3
25	0,1084		P4
16	0,41	SQ3.2	Assurance of wastewater collection from operation
5,26	0,0215		P1
5,26	0,0215		P2
5,26	0,0215		P3
10,53	0,0429		P4
15,79	0,0644		P5
15,79	0,0644		P6
5,26	0,0215		P7
10,53	0,0429		P8
15,79	0,0644		P9
10,53	0,0429		P10
17	0,43	SQ3.3	Supervision and control of the wastewater collection service
16,67	0,0723		P1
16,67	0,0723		P2
33,33	0,1445		P3
16,67	0,0723		P4
16,67	0,0723		P5
30	0,77	SQ3.4	Time taken to resolve incidents in the wastewater collection network
5	0,13	SQ3.5	Time taken to connect to the wastewater service
15	0,38	SQ3.6	Stormweather incidents

SQ4 User service

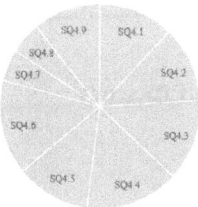

Relative Weight	Absolute Weight	Indicator	
12	0,41	SQ4.1	Complaint management and user satisfaction monitoring
12,5	0,051		P1
12,5	0,051		P2
37,5	0,153		P3
25	0,102		P4
12,5	0,051		P5
12	0,41	SQ4.2	User service quality
14,29	0,0583		P1
14,29	0,0583		P2
14,29	0,0583		P3
14,29	0,0583		P4
14,29	0,0583		P5
14,29	0,0583		P6
14,29	0,0583		P7
13	0,44	SQ4.3	Commitment to user service and contingency information
22,22	0,0982		P1
22,22	0,0982		P2
11,11	0,0491		P3
11,11	0,0491		P4
11,11	0,0491		P5
22,22	0,0982		P6
15	0,51	SQ4.4	Perception of general user satisfaction
12	0,41	SQ4.5	User perceptions of problem resolution quality
15	0,51	SQ4.6	Number of customer service complaints per 100 users and year
5	0,17	SQ4.7	Customer call service waiting time
5	0,17	SQ4.8	Customer service center waiting time
11	0,37	SQ4.9	Time taken to resolve problems

PE Investment Planning and Implementation Efficiency

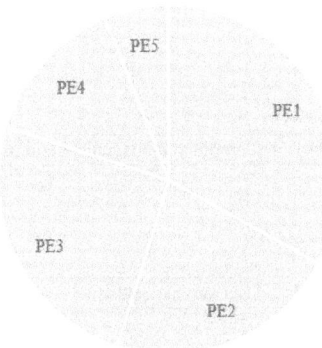

Relative Weight	Absolute Weight	Sub-Area	
33	3,96	PE1	Investment plan content and efficiency
22	2,64	PE2	Investment plan implementation efficiency
25	3	PE3	Existing physical asset management efficiency
14	1,68	PE4	Emergency planning
6	0,72	PE5	Research and development

PE1 Investment plan content and efficiency

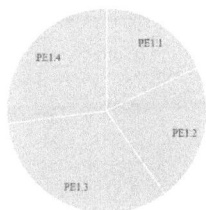

Relative Weight	Absolute Weight	Indicator	
18	0,71	PE1.1	Investment plan contents
8,33	0,0594		P1
8,33	0,0594		P2
8,33	0,0594		P3
8,33	0,0594		P4
8,33	0,0594		P5
8,33	0,0594		P6
8,33	0,0594		P7
8,33	0,0594		P8
8,33	0,0594		P9
25	0,1782		P10
22	0,87	PE1.2	Diagnosis methodology
6,25	0,0545		P1
6,25	0,0545		P2
6,25	0,0545		P3
6,25	0,0545		P4
6,25	0,0545		P5
6,25	0,0545		P6
6,25	0,0545		P7
6,25	0,0545		P8
6,25	0,0545		P9
6,25	0,0545		P10
18.75	0,1634		P11
18,75	0,1634		P12
33	1,31	PE1.3	Methodology for identifying and analyzing alternatives and defining solutions
5,88	0,0769		P1
5,88	0,0769		P2
5,88	0,0769		P3
5,88	0,0769		P4
5,88	0,0769		P5
5,88	0,0769		P6
5,88	0,0769		P7
11,76	0,1537		P8
5,88	0,0769		P9
11,76	0,1537		P10
11,76	0,1537		P11
5,88	0,0769		P12
5,88	0,0769		P13
5,88	0,0769		P14

27	1,07	PE1.4	Methodology for analyzing the plan's financial aspects
25	0,2673		P1
25	0,2673		P2
25	0,2673		P3
25	0,2673		P4

PE2 Investment plan implementation efficiency

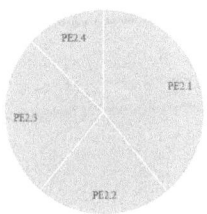

Relative Weight	Absolute Weight	Indicator	
39	1,03	PE2.1	Systems for monitoring implementation of investment plan projects
7,14	0,0735		P1
7,14	0,0735		P2
7,14	0,0735		P3
21,43	0,2206		P4
21,43	0,2206		P5
21,43	0,2206		P6
14,29	0,1471		P7
22	0,58	PE2.2	Compliance with the investment plan
26	0,69	PE2.3	Degree of cost variation in completed works
20	0,1373		P1
20	0,1373		P2
30	0,2059		P3
30	0,2059		P4
13	0,34	PE2.4	Degree of deviation from deadlines established for implementation of works
20	0,0686		P1
20	0,0686		P2
30	0,103		P3
30	0,103		P4

PE3 Existing physical asset management efficiency

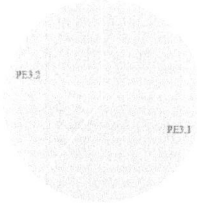

Relative Weight	Absolute Weight	Indicator	
61	1,83	PE3.1	Physical asset management
27,27	0,4991		P1
45,45	0,8318		P2
9,09	0,1664		P3
9,09	0,1664		P4
9,09	0,1664		P5
39	1,17	PE3.2	Annual investment in replacement of fixed physical assets

PE4 Emergency planning

Relative Weight	Absolute Weight	Indicator	
100	1,68	PE4.1	Emergency plan
14,29	0,24		P1
14,29	0,24		P2
14,29	0,24		P3
14,29	0,24		P4
14,29	0,24		P5
14,29	0,24		P6
14,29	0,24		P7

PE5 Research and development

Relative Weight	Absolute Weight	Indicator	
71	0,51	PE5.1	Research and development
31,25	0,1598		P1
6,25	0,032		P2
6,25	0,032		P3
6,25	0,032		P4
6,25	0,032		P5
18,75	0,0959		P6
6,25	0,032		P7
6,25	0,032		P8
12,5	0,0639		P9
29	0,21	PE5.2	Investment in research and development

OE Operating Efficiency

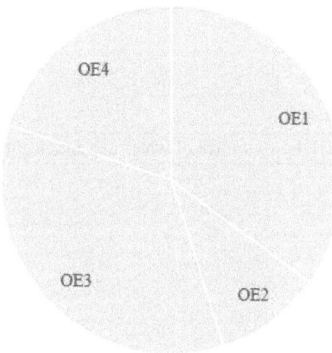

Relative Weight	Absolute Weight	Sub-Area	
35	4,55	OE1	Water resource management efficiency
10	1,3	OE2	Energy usage efficiency
35	4,55	OE3	Infrastructure management efficiency
20	2,6	OE4	Operational and maintenance cost efficiency

OE1 Water resource management efficiency

Relative Weight	Absolute Weight	Indicator	
15	0,68	OE1.1	Control of water use and destinations
14,29	0,0975		P1
21,43	0,1463		P2
7,14	0,0488		P3
7,14	0,0488		P4
7,14	0,0488		P5
7,14	0,0488		P6
14,29	0,0975		P7
7,14	0,0488		P8
7,14	0,0488		P9
7,14	0,0488		P10
20	0,91	OE1.2	Control of water at points of use and consumption
18	0,82	OE1.3	Management of real losses
7,14	0,0585		P1
21,43	0,1755		P2
7,14	0,0585		P3
14,29	0,117		P4
14,29	0,117		P5
14,29	0,117		P6
7,14	0,0585		P7
7,14	0,0585		P8
7,14	0,0585		P9
25	1,14	OE1.4	Real losses in the water supply, transportation and distribution infrastructure
7	0,32	OE1.5	Management of water used in operation
16,67	0,0531		P1
16,67	0,0531		P2
16,67	0,0531		P3
33,33	0,1062		P4
16,67	0,0531		P5
6	0,27	OE1.6	Water used in operation
5	0,23	OE1.7	Management of reclaimed water
20	0,0455		P1
20	0,0455		P2
20	0,0455		P3
20	0,0455		P4
20	0,0455		P5
4	0,18	OE1.8	Reused water

OE2 Energy usage efficiency

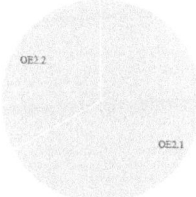

Relative Weight	Absolute Weight	Indicator	
67	0,87	OE2.1	Energy usage efficiency
23,08	0,201		P1
23,08	0,201		P2
15,38	0,134		P3
15,38	0,134		P4
7,69	0,067		P5
15,38	0,134		P6
33	0,43	OE2.2	Energy use in reducing pollutant load

OE3 Infrastructure management efficiency

Relative Weight	Absolute Weight	Indicator	
20	0,91	OE3.1	Efficiency in management of water intake, treatment and distribution infrastructure
6,25	0,0569		P1
6,25	0,0569		P2
6,25	0,0569		P3
6,25	0,0569		P4
12,5	0,1138		P5
12,5	0,1138		P6
12,5	0,1138		P7
18,75	0,1706		P8
6,25	0,0569		P9
6,25	0,0569		P10
6,25	0,0569		P11
14	0,64	OE3.2	Number of ruptures in transportation and distribution pipes
14	0,64	OE3.3	Number of ruptures in service connections (connections up to private supply systems)
8	0,36	OE3.4	Expenditure on corrective maintenance of fixed physical assets linked to the water intake, treatment and distribution system
10	0,46	OE3.5	Expenditure on preventive maintenance of fixed physical assets linked to the water intake, treatment and distribution system
15	0,68	OE3.6	Efficiency in management of wastewater collection and treatment infrastructure
7,14	0,0488		P1
7,14	0,0488		P2
7,14	0,0488		P3
7,14	0,0488		P4
14,29	0,0975		P5
14,29	0,0975		P6
21,43	0,1463		P7
7,14	0,0488		P8
7,14	0,0488		P9
7,14	0,0488		P10
7	0,32	OE3.7	Fortuitous incidents affecting the wastewater collection network during dry weather

5	0,23	OE3.8	Expenditure on corrective maintenance of fixed physical assets linked to the wastewater collection and treatment system.
7	0,32	OE3.9	Expenditure on preventive maintenance of fixed physical assets linked to the wastewater collection and treatment system

OE4 Operational and maintenance cost efficiency

OE4.1

Relative Weight	Absolute Weight	Indicator	
100	2,6	OE4.1	Operational and maintenance cost efficiency
7,69	0,2		P1
7,69	0,2		P2
15,38	0,4		P3
15,38	0,4		P4
15,38	0,4		P5
15,38	0,4		P6
23,08	0,6		P7

ME Business Management Efficiency

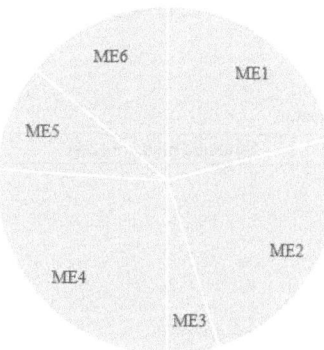

Relative Weight	Absolute Weight	Sub-Area	
21	1,68	ME1	Strategic planning
24	1,92	ME2	Management control
5	0,4	ME3	Organizational structure
26	2,08	ME4	Human resource management
10	0,8	ME5	Procurement management
14	1,12	ME6	Staff and support resource efficiency

ME1 Strategic planning

Relative Weight	Absolute Weight	Indicator	
30	0,5	ME1.1	Strategic plan contents
4,35	0,0219		P1
4,35	0,0219		P2
8,7	0,0438		P3
4,35	0,0219		P4
17,39	0,0877		P5
13,04	0,0657		P6
21,74	0,1096		P7
8,7	0,0438		P8
13,04	0,0657		P9
4,35	0,0219		P10
70	1,18	ME1.2	Strategic plan formulation and implementation
13,33	0,1568		P1
6,67	0,0784		P2
6,67	0,0784		P3
6,67	0,0784		P4
20	0,2352		P5
6,67	0,0784		P6
13,33	0,1568		P7
20	0,2352		P8
6,67	0,0784		P9

ME2 Management control

Relative Weight	Absolute Weight	Indicator	
100	1,92	ME2.1	Management control system
20	0,384		P1
6,67	0,128		P2
13,33	0,256		P3
20	0,384		P4
6,67	0,128		P5
13,33	0,256		P6
20	0,384		P7

ME3 Organizational structure

ME3.1

Relative Weight	Absolute Weight	Indicator	
100	0,4	ME3.1	Organizational structure
8,33	0,0333		P1
16,67	0,0667		P2
16,67	0,0667		P3
25	0,1		P4
16,67	0,0667		P5
8,33	0,0333		P6
8,33	0,0333		P7

ME4 Human resource management

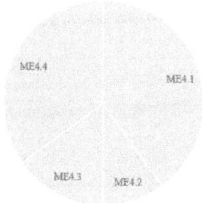

Relative Weight	Absolute Weight	Indicator	
40	0,83	ME4.1	Human resource management
13,04	0,1085	P1	
8,7	0,0723	P2	
8,7	0,0723	P3	
4,35	0,0362	P4	
13,04	0,1085	P5	
8,7	0,0723	P6	
4,35	0,0362	P7	
13,04	0,1085	P8	
4,35	0,0362	P9	
8,7	0,0723	P10	
4,35	0,0362	P11	
8,7	0,0723	P12	
10	0,21	ME4.2	Staff recruited competitively
15	0,31	ME4.3	Staff receiving training
35	0,73	ME4.4	Staff who comply with key position job descriptions

ME5 Procurement management

Relative Weight	Absolute Weight	Indicator	
39	0,31	ME5.1	Procurement
11,76	0,0367		P1
11,76	0,0367		P2
5,88	0,0184		P3
17,65	0,0551		P4
11,76	0,0367		P5
17,65	0,0551		P6
5,88	0,0184		P7
5,88	0,0184		P8
5,88	0,0184		P9
5,88	0,0184		P10
23	0,18	ME5.2	Purchases made by public tender
23	0,18	ME5.3	Successful tenders
15	0,12	ME5.4	Tenders held within the regulated minimum timeframe

ME6 Staff and support resource efficiency

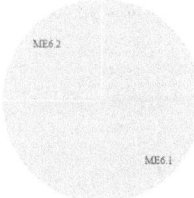

Relative Weight	Absolute Weight	Indicator	
75	0,84	ME6.1	Staff productivity
25	0,28	ME6.2	Expenditure on management and sales functions

FS Financial Sustainability

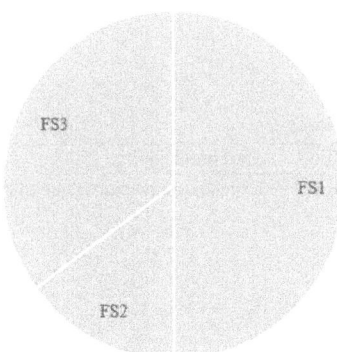

Relative Weight	Absolute Weight	Sub-Area	
50	8,5	FS1	Overall financial sustainability
15	2,55	FS2	Financial management
35	5,95	FS3	Customer management

FS1 Overall financial sustainability

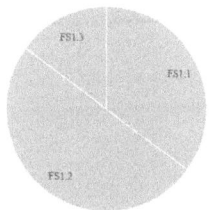

Relative Weight	Absolute Weight	Indicator	
35	2,98	FS1.1	Financial sustainability
33,33	0,9917		P1
16,67	0,4958		P2
6,67	0,1983		P3
10	0,2975		P4
6,67	0,1983		P5
10	0,2975		P6
6,67	0,1983		P7
3,33	0,0992		P8
6,67	0,1983		P9
50	4,25	FS1.2	Expense coverage
30	1,275		P1
20	0,85		P2
30	1,275		P3
20	0,85		P4
15	1,28	FS1.3	Return on equity

FS2 Financial management

Relative Weight	Absolute Weight	Indicator	
30	0,77	FS2.1	Financing, risk hedging and internal control
11,11	0,085		P1
11,11	0,085		P2
11,11	0,085		P3
11,11	0,085		P4
11,11	0,085		P5
22,22	0,17		P6
22,22	0,17		P7
20	0,51	FS2.2	Current ratio
15	0,38	FS2.3	Debt-to-equity ratio
15	0,38	FS2.4	Committed flows
10	0,26	FS2.5	Currency risk
10	0,26	FS2.6	Rate risk

375

FS3 Customer management

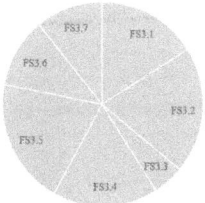

Relative Weight	Absolute Weight	Indicator	
16	0,95	FS3.1	Billing and collection
20	0,1904		P1
6,67	0,0635		P2
6,67	0,0635		P3
13,33	0,1269		P4
13,33	0,1269		P5
6,67	0,0635		P6
6,67	0,0635		P7
13,33	0,1269		P8
13,33	0,1269		P9
20	1,19	FS3.2	Billing effectiveness
5	0,3	FS3.3	Billing error rate
17	1,01	FS3.4	Unbilled water
20	1,19	FS3.5	Collection rate
11	0,65	FS3.6	Average collection time
11	0,65	FS3.7	Arrearage

AS Access to Service

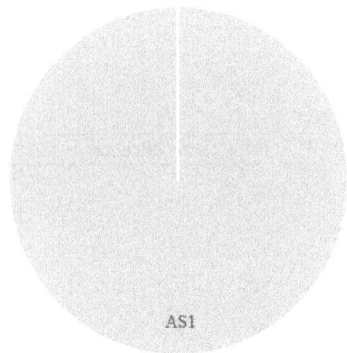

Relative Weight	Absolute Weight	Sub-Area	
100	11	AS1	Access to service

AS1 Access to service

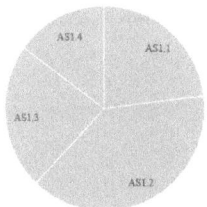

Relative Weight	Absolute Weight	Indicator	
23	2,53	AS1.1	Guaranteed access to service
20	0,506		P1
20	0,506		P2
20	0,506		P3
6,67	0,1687		P4
20	0,506		P5
13,33	0,3373		P6
39	4,29	AS1.2	Household access to drinking water
23	2,53	AS1.3	Connection to wastewater collection systems
15	1,65	AS1.4	Household ability to pay for services received

CG Corporate Governance

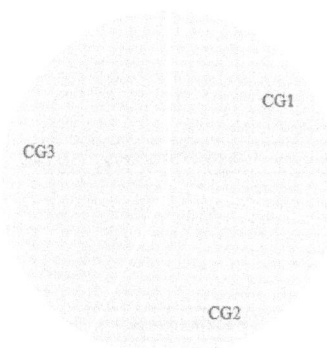

Relative Weight	Absolute Weight	Sub-Area	
29	2,03	CG1	Utility autonomy and responsibilities
29	2,03	CG2	Decision-making processes and accountability
42	2,94	CG3	Transparency and control

CG1 Utility autonomy and responsibilities

Relative Weight	Absolute Weight	Indicator	
100	2,03	CG1.1	Utility autonomy and responsibilities
14,29	0,29		P1
14,29	0,29		P2
14,29	0,29		P3
14,29	0,29		P4
14,29	0,29		P5
28,57	0,58		P6

CG2 Decision-making processes and accountability

Relative Weight	Absolute Weight	Indicator	
35	0,71	CG2.1	Corporate governance
15,38	0,1093		P1
15,38	0,1093		P2
7,69	0,0547		P3
7,69	0,0547		P4
7,69	0,0547		P5
7,69	0,0547		P6
7,69	0,0547		P7
7,69	0,0547		P8
7,69	0,0547		P9
7,69	0,0547		P10
7,69	0,0547		P11
35	0,71	CG2.2	Selection of board members and chief executive officer
20	0,1421		P1
20	0,1421		P2
20	0,1421		P3
10	0,0711		P4
10	0,0711		P5
10	0,0711		P6
10	0,0711		P7
30	0,61	CG2.3	Board of directors' powers and responsibilities
11,11	0,0677		P1
11,11	0,0677		P2
11,11	0,0677		P3
11,11	0,0677		P4
11,11	0,0677		P5
11,11	0,0677		P6
22,22	0,1353		P7
11,11	0,0677		P8

CG3 Transparency and control

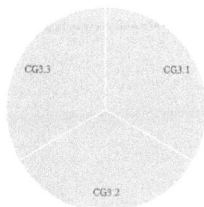

Relative Weight	Absolute Weight	Indicator	
33,33	0,98	CG3.1	Disclosure of information about service delivery
16,67	0,1633		P1
16,67	0,1633		P2
16,67	0,1633		P3
16,67	0,1633		P4
16,67	0,1633		P5
16,67	0,1633		P6
33,33	0,98	CG3.2	Disclosure of institutional and financial information
10	0,098		P1
10	0,098		P2
30	0,294		P3
10	0,098		P4
10	0,098		P5
10	0,098		P6
10	0,098		P7
10	0,098		P8
33,33	0,98	CG3.3	Auditing and control processes
25	0,245		P1
25	0,245		P2
25	0,245		P3
25	0,245		P4

ES Environmental Sustainability

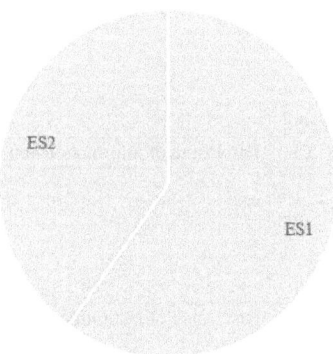

Relative Weight	Absolute Weight	Sub-Area	
60	9	ES1	Wastewater treatment and management
40	6	ES2	Environmental management

ES1 Wastewater treatment and management

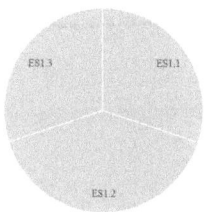

Relative Weight	Absolute Weight	Indicator	
30	2,7	ES1.1	Assurance of operation and control of wastewater treatment services
9,09	0,2455		P1
13,64	0,3682		P2
4,55	0,1227		P3
13,64	0,3682		P4
13,64	0,3682		P5
9,09	0,2455		P6
13,64	0,3682		P7
4,55	0,1227		P8
9,09	0,2455		P9
4,55	0,1227		P10
4,55	0,1227		P11
40	3,6	ES1.2	Availability of operational wastewater treatment infrastructure
30	2,7	ES1.3	Degree of compliance with discharge regulations

ES2 Environmental management

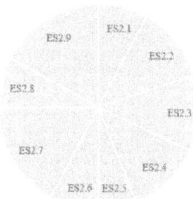

Relative Weight	Absolute Weight	Indicator	
8	0,48	ES2.1	Environmental management framework
16,67	0,08		P1
33,33	0,16		P2
50	0,24		P3
12	0,72	ES2.2	Environmental implications in planning
17,65	0,1271		P1
11,76	0,0847		P2
11,76	0,0847		P3
11,76	0,0847		P4
11,76	0,0847		P5
11,76	0,0847		P6
5,88	0,0424		P7
17,65	0,1271		P8
12	0,72	ES2.3	Environmental operation and promotion
15,38	0,1108		P1
15,38	0,1108		P2
15,38	0,1108		P3
7,69	0,0554		P4
15,38	0,1108		P5
15,38	0,1108		P6
7,69	0,0554		P7
7,69	0,0554		P8
12	0,72	ES2.4	Water withdrawal in relation to the renewable resource
6	0,36	ES2.5	Energy consumption balance
8	0,48	ES2.6	Greenhouse gas emissions linked to drinking water and/or wastewater management
17	1,02	ES2.7	Environmental management of sludge produced by treatment processes
8	0,48	ES2.8	Water resource use
17	1,02	ES2.9	Compliance with environmental regulations

Lightning Source UK Ltd.
Milton Keynes UK
UKOW06f0317100816

280279UK00007B/149/P